Lecture Notes
in Economics and
Mathematical Systems

Operations Research, Computer Science, Social Science

Edited by M. Beckmann, Providence, G. Goos, Karlsruhe, and
H. P. Künzi, Zürich

72

T. P. Bagchi
J. G. C. Templeton

Numerical Methods in Markov
Chains and Bulk Queues

Springer-Verlag
Berlin · Heidelberg · New York 1972

Dr. Tapan P. Bagchi
Prof. James G. C. Templeton

Department of Industrial Engineering
University of Toronto
Toronto, Canada

AMS Subject Classifications (1970): 60 K 25, 90-04, 65 Q 05

ISBN 3-540-05996-2 Springer-Verlag Berlin · Heidelberg · New York
ISBN 0-387-05996-2 Springer-Verlag New York · Heidelberg · Berlin

Offsetdruck: Julius Beltz, Hemsbach/Bergstr.

Dedicated to the memory of
the late Basudev Prasad Bagchi,
father of Tapan Prasad Bagchi.

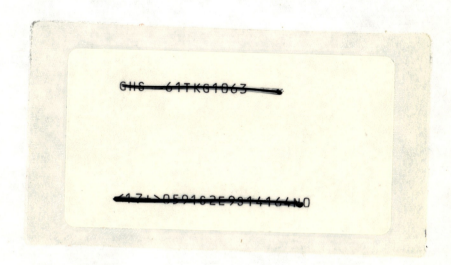

ACKNOWLEDGEMENTS

The authors wish to express their deep gratitude to Professor Ben Bernholtz, who provided the valuable support and encouragement needed for the completion of this work. They are also indebted to Professor M.J.M. Posner, who helped them with valuable discussions and criticism of an earlier version of some of this work. They gratefully acknowledge the financial support of the National Research Council of Canada, and the School of Graduate Studies, University of Toronto, for its University of Toronto Open Fellowship award. Finally, they wish to thank Mrs. Frances Mitchell, who did an excellent job of typing the manuscript with speed and accuracy.

CONTENTS

The following symbolism is based on the original taxonomy of
waiting lines due to Kendall.

M/G/1   a simple (single arrival, single service) single server
          queue with Poisson input, general service times and un-
          limited waiting space.

GI/M/1   a simple single server queue with general independent
          interarrival times, exponential service times and unlimited
          waiting space.

M/M/1,K   a simple single server queue with Poisson input, exponen-
          tial service times and a waiting room that allows a maximum
          of K customers to be present in the system at any time.

$M^X/G^Y/1$ a single server bulk queue with compound Poisson input,
          i.i.d. service times together with i.i.d. service capaci-
          ties for successive services (general batch service) and
          unlimited waiting space.

$M^X/G^Y/1,K$ a single server bulk queue with compound Poisson input,
          general batch service and a waiting room that allows a
          maximum of K customers to be present in the system at any
          time.

Listed below are frequently employed symbols, followed by a statement of their meaning.

$(\cdot)^+$ $\max(0,\cdot)$,   as in $X^+ = \max(0, X)$.

$(\cdot)^-$ $\min(0,\cdot)$.

$\{x,y,z,\ldots\}$ the set with elements x, y, z, $\ldots$.

ϵ 'belongs to' .

M the algebra of all signed measures on Borel subsets of R .

$\Pi(\theta)$ the Wendel operator acting on the elements of M , employed to sweep probability mass in the manner induced by the measurable function $(\cdot)^+ ((\cdot)^-)$.

$\int_{D_\xi} (\cdot)\, d\xi$ a contour integral where D_ξ is a circle in the complex ξ-plane with centre at $\xi = 0$ and radius $|\xi|$.

$(a, b]$ an interval open at left and closed at right.

t time.

$t_0, t_1, t_2, \ldots, t_n, \ldots$ epochs marked by the temporal parameter $n = 0, 1, 2,$ etc.

$t_n - 0 \; (t_n + 0)$ an epoch immediately preceding (following) the epoch t_n .

Q_n number of customers in a queuing system at an appropriately defined instant marked by n .

$X_n (Y_n)$ appropriately defined number of customers arriving (departing).

Prob [·]   the probability that event [·] occurs.

$\lambda(\mu)$ the reciprocal of the mean of exponentially distributed interarrival (service) times.

$\lambda_n (\mu_n)$ the reciprocal of the mean of exponentially distributed interarrival (service) times during a period marked by n .

$Y(s)$ the probability generating function (p.g.f.) of a lattice distributed random variable Y .

$F_\mu (s)$ the p.g.f. associated with the probability measure μ of a lattice distribution.

CHAPTER I

INTRODUCTION

1.1 A PERSPECTIVE

Queues, waiting lines, congestion phenomena, stochastic service systems occur numerously both in the real world and in the professional literature. Unfortunately however, most results from the theory of queues find only limited popularity among those concerned with the understanding, design and control of real life queuing systems. Bhat [1969], Saaty [1966] and Lee [1966], among others, have discussed the nature of and the reasons for the gap between queuing theory and the handling of queuing problems in the real world.

In their attempt to bridge the gap by applying theory to practice, practitioners of operational research have employed a number of techniques. Some of these are listed below.

1) The use of simplified models, for which solutions are available, in place of more accurate but less accessible models. Assuming exponential distributions and steady state solutions may (or may not) lead to a result sufficiently accurate for the problem at hand. Mack [1967] discusses some pitfalls in this procedure.

2) The use of approximate models, and in particular models with a continuous state space, in which the discreteness of the number of customers in the system is neglected. If the number of customers in the system is large, the error in the model may be proportionately small (Bailey [1964], Newell [1971]). Also useful as approximations are the results of heavy traffic theory. See for example Iglehart and Whitt [1970a, 1970b].

3)   The use of inequalities to set upper and lower bounds for quantities of interest. These inequalities may be analytical, as given by Kingman [1970] and Kleinrock, Muntz and Hsu [1971] among others. Numerical inequalities may also be obtained directly from tabulated solutions in some cases, as noted by Lee [1966]. Lee gives bounds for queues with complicated queue disciplines in terms of graphical solutions for queues with service in order of arrival, and for queues with service in random order.

4)   Simulation of queuing models to obtain estimates of quantities of interest in a particular application. Because of the heavy computations which may be needed to obtain sufficiently reliable results, simulation is often regarded as a last resort, to be undertaken only after all other approaches have been exhausted. Simulation is, nevertheless, often useful and sometimes indispensable.

5)   Numerical calculation of probabilities. This approach, although not new, seems to have received less study than its potential usefulness deserves. Some previously published methods for numerical calculation of state probabilities are mentioned in Section 1.2 below.

The present study is mainly devoted to a new numerical method of computing transient state probabilities for the imbedded Markov chains of queue length processes. An essentially abstract formalism to handle Markovian relationships is set forth as an extension of the algebraic formalism of queuing problems proposed by Kingman [1966]. This formalism is utilized to develop a numerical procedure that yields the exact time-dependent state probabilities. A variety of practically significant problems are treated by this method. Some analytical approximations that relate to queue length processes and are suitable for numerical work are also given.

Problems of specification of queuing models and estimation of

parameters are not considered in the present treatment. Also omitted
are problems of optimal design and operation of waiting line systems.

1.2 EARLIER WORK

Over the past two decades steady progress has been made toward
solving increasingly difficult and realistic waiting line models.
Nevertheless, only a small proportion of the new "analytical" results
thus obtained has been found suitable for *numerical* work: a desirable
feature when one has to compare tradeoffs among various decision
alternatives. This alone has perpetuated much of the general belief
that simulation is the only approach likely to yield a practical
answer in the study of complex waiting line systems. The lack of results suited for ready practical implementation is observed in several
areas in waiting line theory. One such class of waiting line models is
distinguished by the presence of a specific feature, namely that customers arrive in groups of random size, and are served in groups that
are themselves of random size.

Waiting line models belonging to the above category are termed
"bulk queues" in the literature. Families arriving at a restaurant,
elevators serving a floor, children awaiting their turn at a roller
coaster or platoons of cars moving through a street intersection at a
change of light form examples of the group arrival, group service
phenomena. This feature of treatment of customers or entities in
batches, to replace the more basic structure of single arrival and
single service, broadens the applicability of queuing models. It also
promotes an improved understanding of the behaviour of storage systems,
inventories and dams. Despite their highly relevant character,
explicit results available for bulk queuing systems that are convenient
for numerical evaluation are relatively few. An overview of the investigations conducted on such systems appears below.

A dearth of usable analytical results is also observed for another class of waiting line models, characterized by the presence of non-stationary arrival and servicing conditions. The few investigations that have probed this area will be referenced in the sequel.

Bailey [1954] carried out the first mathematical investigation of queues involving batch service. He studied the stationary behaviour of a single server queue having simple Poisson input, intermittently available server and service in batches of a fixed maximum size. The results of this study are given in terms of a probability generating function, the evaluation of which requires determining the zeros of a certain polynomial. This study was followed by a series of papers involving the treatment of queuing processes with group arrival or batch service. Saaty's "Elements of Queueing Theory" contains a good account of some of these works. Miller [1959] was the first investigation to treat the general problem in which customers arrive in groups and are served in groups. Miller examines the problem with an imbedded Markov chain to obtain stationary distributions in the $M^X/G^Y/1$ queue (the notation implying compound Poisson arrivals and general batch service with a single server), and in the $GI^X/M^Y/1$ case — a bulk queue with general input, single server and exponential server times. Miller's results, like Bailey's, necessitate the evaluation of the zeros of certain polynomials, in order that the stationary state probabilities of the queues considered may be evaluated.

Keilson [1962] discusses the queue length process of the $M^X/G^Y/1$ queue by first using the supplementary variable method (Cox [1955]) to obtain a partial differential equation. He then transforms this equation to a boundary problem on the unit circle, of the type described by Muskhelishvili [1953] as a non-homogeneous Hilbert problem. Solution of the Hilbert problem gives the Laplace transform of the probability generating function.

Bhat [1964] has used fluctuation theory to analyse the imbedded Markov chains of two single server bulk queues, namely, the $M^X/G^Y/1$ and the $GI^X/M^Y/1$ systems. He studies the equilibrium behaviour of these queues and expresses his results in terms of a celebrated result in probability theory: Spitzer's identity. Notwithstanding the compact appearance of these results, computational considerations reveal quickly that the "Spitzer's identity" form is not particularly suited for an easy evaluation when numerical results are desired. A similar observation has been made by Kingman [1965] with relation to the distribution of waiting times in the GI/G/1 queue. Teghem, Loris-Teghem and Lambotte [1969] consider the $M^X/G^Y/1$ and $GI^X/M^Y/1$ queuing systems using the theory of semi-Markov processes. The final evaluation of their results requires evaluation of polynomial roots. Cohen [1969] presents an elegant and elaborate analysis of the transient and stationary behaviour of the $M^X/G^Y/1$ queue utilizing a very powerful analytical method based on the Pollaczek approach. A recent (unpublished) investigation by Chaudhry and Templeton uses the Erlangian (difference-differential equation) approach to obtain results for a bulk queue with compound Poisson input and mixed Erlang service times. A distinctive feature of this study is that it obtains as special cases a number of results known from earlier investigations. The final results of this study, nevertheless, require invoking Rouché's theorem for their evaluation.

An interesting variant of the studies mentioned above, all of which have invariably assumed that the size of the waiting area available to the customers is unlimited, is Kinney's [1962] investigation. Kinney considers a discrete time bulk queue with a finite sized waiting area. The maximum number of customers allowed in the system is assumed to be a fixed number, N . Stationary and (discrete time) transient distributions of the number of customers in the system are derived in terms of generating functions as an extension of an absorbing barrier process.

Simple though it might seem, computational experience with analytical results invoking, for instance, Rouché's theorem asserts that it may be sometimes rather difficult to execute the pertinent manipulations involving zeros of polynomials, even on a computer, particularly when the problem is ill-conditioned. This fact stands somewhat ignored with relation to a variety of queuing systems, and in particular bulk queues that are claimed to be "solved" analytically. A good discussion on the subject is found in a paper by Page [1965].

Turning to the non-stationary character of real life waiting lines, one observes that in many situations the input and servicing conditions alter with time. This is rather typical of traffic systems, or in restaurants where arrivals are more probable at rush hours than at slack periods. The analysis of such systems becomes comparatively involved, as is to be expected. Clarke [1953] and Luchak [1956] have analytically studied simple queues with exponentially distributed interarrival times, and having a mean arrival rate which varies with time. Clarke suggests a standard method for obtaining the state probabilities in terms of the solution of a Volterra-type integral equation. Luchak's solution consists of a Taylor's series expansion, which unfortunately presents some computational problems. An elaborate discussion on the computational performance of Clarke's and Luchak's, and several other related methods, is available in Leese [1964] and Leese and Boyd [1966]. The different numerical method suggested in Leese [1964] and Leese and Boyd [1966], one may note, involves an approximate numerical solution of a certain integral equation and applies only to a simple Markovian queue.

Hasofer [1964] has considered a single server queue with non-homogeneous Poisson input and general service times. His discussion concentrates mainly on the case when the parameter of the Poisson input $\lambda(t)$ is a periodic function of time. The results of this study contain explicit forms only for the asymptotic values for the empty system probability and the Laplace transform of the waiting time distribution, although the analysis itself is remarkably elaborate.

Kerridge [1966] gives a method for numerically evaluating the state probabilities in queues involving exponential, and in the special "single service" case the Erlangian distribution. His method involves matrix operations and applies to finite or (truncated) infinite queues. No indication is unfortunately given in his paper of the effects of round-off errors, on the computed state probabilities. A numerical method is also given by Satyamurty [1965] to evaluate transient results for a finite waiting room M/M/1 queue, where customers are permitted to balk. The method comprises a procedure involving the eigenvalues of a certain matrix, and appears satisfactory for a queue with a small waiting room. Hillier and Boling [1967] present an exact and an approximate numerical procedure to obtain the steady state performance of a system consisting of a number of finite queues in series. They utilize the basic properties of the exponential and the Erlang distributions to set up the pertinent transition matrices. A good indication of the computational accuracy of the procedures involved is given in their paper.

Hirasawa [1971] obtains numerical solutions to first-come-first-served fixed batch service queues. His analysis proceeds by considering the underlying imbedded Markov chain and yields an iterative method by which the steady state probabilities may be calculated. He also presents numerical methods to find the steady state mean waiting times for several exponential and Erlangian queues. Typical numerical results are tabulated and a graphical comparison of the exact and some approximate solutions is given. Little indication is given, however, of the magnitude of computational errors that would be committed by employing the prescribed numerical procedures. Neuts [1971] has numerically studied a class of finite queues in discrete time. This class of queues has been proposed as an alternative to the classical $M/G/1$ and $GI/M/1$ type continuous time queuing models. A recurrence method is suggested to evaluate the transient state probabilities, along with a procedure to evaluate the transient distribution of the integer valued waiting times. A FORTRAN IV program to perform the necessary operations is given in this work; it would be of interest to learn about the numerical accuracy of the algorithms employed here. A set of examples is given to show that this numerical procedure is considerably faster than the approach of simulating the queue. Klimko and Neuts [1972] have treated at length the numerical problems arising in the computation of higher moments of the busy period for certain classical queues of the $M/G/1$ type. They describe a method that is based on the classical differentiation formula due to Fáa di Bruno.

In the next chapter we set the groundwork to expound an efficient numerical procedure for studying the queue length processes in a sizable class of waiting line systems.

CHAPTER II

MATHEMATICAL FORMULATION OF THE BULK QUEUING PROBLEM

2.1 A CLASS OF MARKOV RECURRENCE RELATIONS

Ever since its first exposition by Kendall [1951], the concept of imbedded Markov chain has been widely applied to the analysis of a remarkably large number of queuing and storage processes that are not in general Markovian. Methods of accomplishing this for bulk queues will be discussed in this chapter. Also, an abstract formalism which gives a rigorous and convenient base for conducting a study of such Markov processes will be presented.

Suppose that customers arrive in batches of random size and are served by a single server in batches of random size, not necessarily identical with the arriving batch sizes. Let the interarrival times of the incoming batches be independent and identically distributed random variables. Service times are also independent random variables, with identical proper distributions. Now consider the two following processes.

Process A: Assume that the queuing system is operating in such a way that the server never relaxes. This is to say that the server begins a service period immediately after the completion of the preceding period. Also, at the start of each period he accepts an allowable number of customers. This allowable number is determined by his current capacity for service (e.g. the number of standing places unoccupied in an elevator when the door opens) if there are any customers waiting, or else it is zero if there is no one waiting. Situations resembling the operation of an elevator or street cars picking up passengers at a particular point may be suitably modeled in this manner. The operational rules adopted here typify an 'intermittently available server', in that arrivals joining an empty system must wait for the commencement of the most imminent servicing cycle.

Let t_0 , t_1 , t_2 , ... denote the epochs on the time axis at which the server becomes successively available. Obviously therefore, since the server never relaxes at the end of a servicing period, $t_1 - t_0$, $t_2 - t_1$ etc. are random variables distributed as the servicing times. Let the number of customers in the system at $t_n + 0$ inclusive of those in service be denoted by Q_n . Let the capacity for service during the service period commencing at $t_n + 0$ be given by the random variable Y_n . If now X_{n+1} denotes the total number of customers, a random variable, arriving in the interval $(t_n, t_{n+1}]$ who await the commencement of the next cycle, then it may be shown that the sequence of random variables $\{Q_n, n = 0,1,2,...\}$ satisfies the recurrence relation

$$Q_{n+1} = (Q_n - Y_n)^+ + X_{n+1} , \quad n = 0,1,2,... \qquad (2.1)$$

where $Z^+ = \max(Z, 0)$.

If $\{X_{n+1}\}$ and $\{Y_n\}$ in the above are sequences of independently distributed random variables, then the sequence $\{Q_n, n = 0,1,2,...\}$ will constitute a first order Markov chain in a countable state space, taking values on non-negative integers.

Process B: Changing the picture slightly, we may regard epochs t_0 , t_1 , t_2 , ... as instants when batches, numbered as 0, 1, 2, ... arrive for service. Let Q_n denote the number of customers in the system, including those in service at $t_n - 0$. The server in this case may, out of necessity, remain idle (inactive) when no customers are available. Let X_n (a random variable) denote the size of the batch arriving at t_n, and let Y_n denote the sum of the randomly realized capacities of the server which become available in $[t_{n-1}, t_n)$. Also, assume that latecomers are allowed to join a batch in service to fill up the server's current capacity without affecting his service time. As an example one may consider a movie theatre, which has the house cleared after each showing; customers arriving at 7.30 can see the remainder of the 7.00 showing, as opposed to waiting and seeing all of the 9.15 showing. It may then be shown, assuming that $\{X_n\}$ and $\{Y_n\}$ are sequences of independent random variables, that as for Process A, the sequence $\{Q_n, n = 0,1,2,...\}$ is a first order Markov chain determined by the recurrence relation

$$Q_{n+1} = (Q_n + X_n - Y_{n+1})^+ , \quad n = 0,1,2,... . \tag{2.2}$$

Analysis of infinite waiting space bulk queues can be based on (2.1) or (2.2). These formulations appear to have been used in the work of Bhat [1964] for the first time. With a simple rewording of these formulations in which we regard arriving customers as incoming goods or rain water, and batch services as supplies or releases, the same equations may be used to describe the behaviour of inventories or dams with infinite capacity.

Physical and economic considerations reveal that many realistic waiting line systems have only a finite capacity. Under such conditions, as long as the system, including the waiting and service areas, does not become full the arrivals get accepted. Otherwise, they are turned away because of non-availability of space. A restaurant with limited waiting room makes a practical example of this. Ghosal [1970] has proposed for the $M^X/G^Y/1,K$ queue the recurrence relation

$$Q_{n+1} = \min\left[K, (Q_n + X_n - Y_{n+1})^+\right] , \quad n = 0,1,2,... \tag{2.3}$$

where K is an upper bound on Q_n and observations are taken an instant before every arrival. A shortcoming of this model is that it
allows the actual queue size between two successive observation
epochs to exceed K , violating thereby the underlying limited waiting space assumption.

The limited waiting space bulk queue may alternatively be
modeled as follows. Let the random variables X_{n+1} and Y_n be defined in the manner of Process A. Let K denote the maximum number
of customers allowed in the system at any time, including those being
currently served, and assume that latecomers do not join a batch in
service. One may then show that if Q_n denotes the total number of
customers in the system at $t_n + 0$ then the sequence of random variables $\{Q_n, n = 0,1,2,...\}$ satisfies

$$Q_{n+1} = K + \left[(Q_n - Y_n)^+ + X_{n+1} - K\right]^- , \quad n = 0,1,2,... \qquad (2.4)$$

where $z^+ = \max(Z, 0)$ and $z^- = \min(0, Z)$. This sequence becomes a
first order finite Markov chain, taking values on integers $0,1,2,...,K$
when the X_{n+1} and Y_n are independent random variables.

Likewise, let random variables X_n and Y_{n+1} denote quantities
as in Process B. Assuming the appropriate independence properties it
may then be shown that $\{Q_n , n = 0,1,2,...\}$ satisfies the recurrence relation

$$Q_{n+1} = \left[K - Y_{n+1} + (Q_n + X_n - K)^-\right]^+ , \quad n = 0,1,2,... \qquad (2.5)$$

which also is a finite Markov chain. Relations (2.4) and (2.5)
might also be conceived as inventory models, with random demands and
probabilistic replenishment.

A Markovian relation which relates the waiting times of success-

ive customers in a simple single server queue was first conceived by
Lindley [1952]. The relation, stated as

$$w_{n+1} = (w_n + u_n)^+ , \quad n = 0,1,2,\ldots \qquad (2.6)$$

with $\{u_n\}$ a sequence of independent, identically distributed random
variables has since then been studied by Smith [1953], Pollaczek
[1957], Spitzer [1956], Baxter [1960], Kingman [1962] and others.

One may show that the study of (2.1) and (2.2) reduces to that of
(2.6), which in turn culminates in Spitzer's identity (see Spitzer,
[1956]). The solution of (2.3) is conjectured by Ghosal [1970] to be
a variation of Spitzer's identity which, however, is erroneous. This
is so because the measurable function $\phi(\cdot)$, defined as $\min[K, (\cdot)^+]$,
does not satisfy the pertinent necessary conditions (equation 114,
Kingman [1966]). Continuous state space versions of (2.4) and (2.5)
are studied by Cohen [1967] and Roes [1970] using complex integral
equations. A special case of (2.4), for the M/G/1,K system — a
limited waiting space single server queue with simple Poisson input
and general single service — is studied by Cohen [1969]. A powerful
numerical approach that handles a wide class of recurrence relations
resembling (2.1), (2.2), (2.4) or (2.5) is set forth in Chapter III.
The concept central to this approach originated in an abstract
approach — one that was obtained by stripping the study of congestion
problems of the obscurities of special analytic devices — a position
suggested by Kingman's [1966] work.

2.2   THE ABSTRACT FORMALISM

Let a probability space (Ω, F, P) be given, where Ω is an
abstract space, F a fixed σ-field of events in Ω and P a com-
plete probability measure on F . The matter of interest in many
stochastic processes, such as a queuing process, often is the deter-

mination of the distribution of random variables $X(t)$, where t is an independent parameter and the $X(t)$ — defined as functions on Ω — are all F-measurable, assuming values in the real space. We say that the process $\{X(t), t \geq 0\}$ has the Markov property if the following is true for all n :

$$\text{Prob}\left[X(t_n, \omega) = i_n | X(t_1, \omega) = i_1, \ldots, X(t_{n-1}, \omega) = i_{n-1}\right]$$

$$= \text{Prob}\left[X(t_n, \omega) = i_n | X(t_{n-1}, \omega) = i_{n-1}\right]$$

where $\omega \in \Omega$, $t_1 < t_2 < \cdots < t_n$ and $\{i_n\}$ belong to the range of $X(t)$.

It will suffice for the purpose of studying the bulk queue to restrict the range of $X(t)$ to the set of all integers. The process $\{X(t), t \geq 0\}$ with this restriction becomes a first order Markov chain, which may be characterized as

$$X(t_n) = \phi\left[X(t_{n-1})\right] \tag{2.7}$$

where $\phi(\cdot)$ is measurable. Wendel [1958] and Kingman [1966] set forth an analytic, rather than the traditional probabilistic approach, to the study of processes like (2.7), even though their study is confined to a special queuing process of the Lindley type (2.6). An extension of their formalism, which uncovers the basic structure of a wider class of congestion phenomena, is shown below.

Let M be the Banach algebra (see Kingman [1966]) of all finite signed measures on the Borel subsets of the real line R . M is a real commutative algebra, the operation of multiplication in which is defined as convolution, such that

$$\lambda(E) = \int\int_{\{(x,\,y);\ x+y\,\epsilon\,E\}} \mu(dx)\,\nu(dy)$$

with $\lambda = \mu\nu$ and μ, ν and $\lambda \,\epsilon\, M$. Suppose that $\phi : R \to R$ is any Borel measurable function. One can define a measure $\mu_1 \,\epsilon\, M$ for any measure $\mu \,\epsilon\, M$ such that

$$\mu_1(E) = \mu\big(\phi^{-1}(E)\big) = \mu\{X;\ \phi(X)\,\epsilon\,E\} .$$

If therefore a random variable X has a distribution μ, then $\phi(X)$ has distribution μ_1. For any such $\phi(\cdot)$ the function which maps μ into μ_1 is linear, and is an operator of the Banach algebra M into itself.

Two special operators which will be needed extensively in our analysis are defined as follows.

Let $\phi(X) = X^+$, with $X^+ = \max(X, 0)$. The associated operator $\mu \to \mu_1$ is denoted by Π, so that

$$\Pi\mu(E) = \mu\{X;\ X^+\,\epsilon\,E\} .$$

Also, Π is a projection:

$$\Pi^2\mu(E) = \Pi\big[\Pi\mu(E)\big]$$

$$= \Pi\big[\mu\{X;\ X^+\,\epsilon\,E\}\big]$$

$$= \mu\{X;\ (X^+)^+\,\epsilon\,E\}$$

$$= \mu\{X;\ X^+\,\epsilon\,E\}$$

$$= \Pi\mu(E) .$$

Let the range and the null space of Π be defined as

$$M_{\pi+} = \{\Pi\mu; \ \mu \in M\}$$

and

$$M_{\pi-} = \{\mu \in M : \pi\mu(E) = 0\} \ .$$

Kingman shows that $M_{\pi+}$ and $M_{\pi-}$ are subalgebras of M, whence Π is a Wendel projection. For any $\mu \in M$, $\mu_{\pi+} = \Pi\mu$ belongs to $M_{\pi+}$, and $\mu_{\pi-} = \mu - \Pi\mu$ to $M_{\pi-}$. Hence $\mu = \mu_{\pi+} + \mu_{\pi-}$, a decomposition which is induced by Π and is unique. In intuitive terms, $M_{\pi+}$ consists exactly of measures concentrated on the half-line $[0, \infty)$. The operator Π sweeps all measures in $(-\infty, 0]$ to the origin. $M_{\pi-}$ consists exactly of measures concentrated on $(-\infty, 0]$ with $\mu(R) = 0$ for $\Pi\mu = 0$ iff $\mu \in M_{\pi-}$.

A similar, but reversed decomposition of a measure $\mu \in M$ is induced by Θ, associated with $\phi(X) = X^-$, with $X^- = \min(X, 0)$. For $\Theta : M \to M$ one has

$$M_{\theta+} = \{\Theta\mu; \ \mu \in M\}$$

and

$$M_{\theta-} = \{\mu \in M : \Theta\mu(E) = 0\}$$

as the range and the null space. $M_{\theta+}$ consists exactly of measures concentrated on the half-line $(-\infty, 0]$. The operator Θ sweeps all measures in $[0, \infty)$ to the origin. $M_{\theta-}$ consists exactly of measures concentrated on $[0, \infty)$ with $\mu(R) = 0$. Like Π, Θ is also a Wendel projection, inducing a decomposition of $\mu \in M$ as $\mu_{\theta+} + \mu_{\theta-}$.

Let $\mu \in M$ be the measure associated with a lattice probability distribution, i.e. one concentrated on the integral multiples of some real number. Let this real number be unity. The action of sweeping of probabilities in the subspaces of R, induced by measurable functions $(\cdot)^+$ or $(\cdot)^-$ and carried out by linear operators Π or Θ, may be actualized relatively easily in $M[[s]]$, the set of power series whose coefficients are probabilities, for lattice distributions. (Note the difference between $M[[s]]$ and Kingman's $M[[x]]$, which is a set of power series whose coefficients are probability measures.) A homomorphic mapping of the signed measure algebra M into the set $M[[s]]$ is executed to this end as follows.

$$S \; : \; M \rightarrow M[[s]] \; : \; \mu \; \longrightarrow \; F_\mu(s) \; .$$

This mapping will be referred to as 'S-homomorphism' in the remainder of this monograph.

Relation (2.4) or (2.5) is studied in the present framework as follows. Consider (2.5). Let the distributions of the random variables $K - Y_{n+1}$ and $X_n - K$ be represented by measures f_{n+1} and h_n in M . If g_n denotes the distribution of Q_n , then utilizing operators Π and Θ one may write from (2.5)

$$g_{n+1} = \Pi\left(f_{n+1} \cdot \Theta\left(g_n \cdot h_n\right)\right) \; . \tag{2.8}$$

This is so because f_{n+1} , $h_n \in M$ (the signed measure Banach algebra); since $g_n \in M_{\pi+}$ and $M_{\pi+} \subset M$ we have $g_n \cdot h_n \in M$. Also, $\Theta\left(g_n \cdot h_n\right) \in M_{\Theta+}$, and $M_{\Theta+} \subset M$, hence $\Theta\left(g_n \cdot h_n\right) \in M$, and so $f_{n+1} \cdot \Theta\left(g_n \cdot h_n\right) \in M$. This gives $\Pi\left(f_{n+1} \cdot \Theta\left(g_n \cdot h_n\right)\right) \in M_{\pi+}$ as desired.

Let the initial queue size be z , $0 \le z \le K$. Applying S - homomorphism to (2.8) one obtains

$$G_{n+1}(s) = \Pi\left(F_{n+1}(s) \cdot \Theta\left(G_n(s) \cdot H_n(s)\right)\right) \; , \quad n = 0,1,2,\ldots \tag{2.9}$$

with $G_0(s) = s^Z$. The time dependent behaviour of the queue size distribution $\{G_n(s)\}$ may, in principle, always be studied from (2.9) for any specified form of non-stationarity present in the arrivals or in the servicing of customers. These non-stationarities may be appropriately modeled by functions $\{F_{n+1}(s)\}$ and $\{H_n(s)\}$.

The abstract setting thus reduces the study of the limited waiting space bulk queue to that of a set of relations like (2.9). The distribution of Q_n may now be obtained, provided the actions of the inherent sweeping operators are actualized.

Analytic equivalents of the operators Π and Θ are of Pollaczek type, given by Cohen [1969] for X an integer as

$$
p^{X^+} = \begin{cases} \dfrac{1}{2\pi i} \displaystyle\int_{D_\xi} \left(\dfrac{1}{\xi-p} - \dfrac{1}{\xi-1}\right) \xi^X \, d\xi \ , & |p| < |\xi| < 1 \ , \\[4ex] 1 - \dfrac{1}{2\pi i} \displaystyle\int_{D_\xi} \left(\dfrac{1}{\xi-1} - \dfrac{1}{\xi-p}\right) \xi^X \, d\xi \ , & |\xi| > |p| \ , \ |\xi| > 1 \ , \end{cases}
$$

and

$$
p^{X^-} = \dfrac{1}{2\pi i} \int_{D_\xi} \left(\dfrac{1}{\xi-1} - \dfrac{1}{\xi-p}\right) \xi^X \, d\xi \ , \qquad 1 < |\xi| < |p| \ . \tag{2.10}
$$

In relation to these equivalents the probability generating functions of X^+ and X^- , formerly regarded as $\Pi X(p)$ and $\Theta X(p)$ are $E\left[p^{X^+}\right]$ and $E\left[p^{X^-}\right]$ respectively. Incorporation of the integration equivalents for the sweeping operators Π and Θ into (2.9) gives an

integral equation in complex variables, the solution of which yields
the desired probabilities. The equation, however, is solvable only
when arrivals and services are homogeneous processes.

The integral equation just mentioned might well be derived
directly from (2.5) utilizing (2.10). However, the rather hideous
development of the abstract formalism set forth here serves a useful
and important purpose. We introduce below a simple concept which
works in the setting of (2.9) and makes possible a direct and exact
numerical evaluation of the time-dependent results.

Let $Y(s) = \sum\limits_{-\infty}^{\infty} \eta_i s^i$ be a probability generating function of a
certain random variable Y, taking values on all integers, and
where the probability that Y equals i is equal to η_i,
$i = \ldots -3, -2, -1, 0, 1, 2, \ldots$, $\sum\limits_{i=-\infty}^{\infty} \eta_i = 1$. The generating
function of Y^+ (the function $(\cdot)^+$ being as defined earlier) will
be given by $\Pi Y(s)$, which we may manifestly put as

$$\Pi Y(s) = \sum_{i=-\infty}^{0} \eta_i + \sum_{i=1}^{\infty} \eta_i s^i .$$

Likewise, the generating function of Y^- may be given by $\Theta Y(s)$,
expressed as

$$\Theta Y(s) = \sum_{i=-\infty}^{-1} \eta_i s^i + \sum_{i=0}^{\infty} \eta_i .$$

The action of the operators on $Y(s)$ in either case thus becomes
equivalent to taking a partial summation of certain series — a process amenable to electronic computation. Suppose then that the
exact functional dependence of $F_{n+1}(s)$ and $H_n(s)$ on n, the
temporal parameter of the queuing process, is known. The evolution
of the process described by (2.5) may then be studied numerically

from (2.9) in a recursive, but exact, manner. We employ this principle extensively in Chapter III.

Alternative approaches to derive recursive relations inter-relating distributions of the various random components of queuing systems are possible. However, some such relations may be shown to be equivalent to ones resembling (2.9) which present the central result in a portmanteau form.

CHAPTER III

A NUMERICAL APPROACH TO WAITING LINE PROBLEMS

3.1 NUMERICAL METHODS IN QUEUING THEORY

Queuing problems formulated by equations (2.4) and (2.5) are, in general, very difficult to solve, particularly when the distributions of X_{n+1} and Y_n are arbitrary functions of the temporal parameter n. One of the few available time-dependent solutions treating model parameters as arbitrary functions of time considers no more than the simple M/M/1 queue (Clarke, [1953]). Approximations, or the use of simulation, sometimes provide tentative answers to those waiting line problems for which the difficulties in direct analysis are formidable. However, approximations may not always be satisfactory, while simulation inter alia suffers substantially from large computational requirements.

Numerical techniques to handle queuing problems have recently made their appearance in the queuing literature. If applicable, these techniques at times prove very useful. The utility of the currently available methods, however, is rather limited (see Leese [1964], Kerridge [1966]). The method described in this chapter attempts to broaden the scope of numerical approaches to queuing problems. It works with probability generating functions of random variables, employing the notion of sweeping of probabilities, and obtains exact time-dependent distributions. Also, it permits the convenient handling of non-stationarities of any specified form prevalent in the arrivals or in the servicing of customers.

3.2   THE BASIC THEORY

Many mathematical questions recur in different guises in various seemingly different situations. The theoretical notions of the present approach can be expressed as follows. The abstract formalism set forth in Chapter II, it will be recalled, focusses on expressing the desired distributions, treated as unknown algebraic elements or pro- bability measures, in terms of the known distributions, rather than studying congestion problems as particular concrete situations. The actual functional relationship among the elements is established once the structure of the basic Markov chain describing the process be- comes known. The procedure employs the basic algebraic operators that constitute the signed measure Banach algebra, and also two Wendel operators, Θ and Π , used to sweep probability mass over certain subspaces of the real line. In particular, when lattice distributions are involved, the actions of the Wendel operators become equivalent to those of Pollaczek-type integral operators, or equivalently to the taking of partial summations of certain series. The last assertion delineates the essence of a numerical approach, as indicated below.

Let $Y(s) = \sum\limits_{i=-\infty}^{\infty} \eta_i \, s^i$ be the probability generating function of a certain random variable Y . The probability that Y equals i is given by η_i , $i \in I$, where I is the set of all integers and $\sum\limits_{i=-\infty}^{\infty} \eta_i = 1$. The p.g.f. of Y^+ (the function $(\cdot)^+$ being as de- fined earlier) will be given by $\Pi Y(s)$, which one may put as

$$\Pi Y(s) = \sum_{i=-\infty}^{0} \eta_i + \sum_{i=1}^{\infty} \eta_i \, s^i \ .$$

Likewise, the p.g.f. of Y^- is given by $\Theta Y(s)$, expressed as

$$\Theta \; Y(s) \; = \; \sum_{i=-\infty}^{-1} \eta_i \; s^i \; + \; \sum_{i=0}^{\infty} \eta_i \; .$$

The action of the operators on $Y(s)$ in either case thus becomes equivalent to taking partial summations of certain series, an operation particularly well carried out by computer. This concept, as we show in the following pages, is applicable to the study of a variety of waiting line problems for which previously available techniques have failed to provide satisfactory answers.

3.3 THE SIMPLE QUEUE WITH LIMITED WAITING ROOM

We begin by considering a simple queuing system (i.e. a queue with single arrival and individual service) with finite waiting room, although a discussion based on one with infinite waiting room is entirely possible. A simple queue with independently distributed interarrival times, negative exponential service times and a finite sized waiting room may be modeled (cf. 2.5) as

$$Q_{n+1} = \left[K - Y_{n+1} + (Q_n + 1 - K)^- \right]^+ , \quad n = 1,2,3,\ldots \qquad (3.1)$$

with $Q_1 = z$, $0 \le z \le K$. Q_n is the number of customers present in the system immediately before the arrival of the nth customer, Y_{n+1} the number of services completed between the arrivals of the nth and $n+1^{th}$ customers and K the maximum number of customers allowed in the system. Let the distributions of random variables Q_n and $K - Y_{n+1}$ be represented by measures g_n and f_{n+1} in the measure algebra M . Then, from (3.1) it follows that

$$g_{n+1} = \Pi \left(f_{n+1} \; \Theta \left(g_n \; \delta_{1-K} \right) \right) , \quad n = 1,2,3,\ldots \qquad (3.2)$$

with $g_1 = \delta_z$ where $\delta_r \in M$ denotes a unit probability mass at the point r on the real line. Application of S-homomorphism to (3.2) yields

$$G_{n+1}(s) = \Pi\left[F_{n+1}(s) \ \Theta\left[G_n(s) \ s^{1-K}\right]\right], \quad n = 1, 2, 3, \ldots \tag{3.3}$$

with $G_1(s) = s^z$. Let

$$G_n(s) = g_{n,0} + g_{n,1} \ s + g_{n,2} \ s^2 + \cdots + g_{n,K} \ s^K.$$

Hence

$$\Theta\left[G_n(s) s^{1-K}\right] = g_{n,0} \ s^{1-K} + g_{n,1} \ s^{2-K} + g_{n,2} \ s^{3-K} + \cdots$$

$$\cdots + g_{n,K-2} \ s^{-1} + g_{n,K-1} + g_{n,K}.$$

Now, denote $F_{n+1}(s)$ by

$$f_{n+1,0} \ s^K + f_{n+1,1} \ s^{K-1} + f_{n+1,2} \ s^{K-2} + \cdots + f_{n+1,K} + f_{n+1,K+1} \ s^{-1} + \cdots .$$

Hence

$$F_{n+1}(s) \ \Theta\left[G_n(s) s^{1-K}\right] = \left(g_{n,K-1} + g_{n,K}\right) f_{n+1,0} \ s^K$$

$$+ \left[f_{n+1,1}\left(g_{n,K-1} + g_{n,K}\right) + f_{n+1,0} \ g_{n,K-2}\right] s^{K-1}$$

$$+ \left[f_{n+1,2}\left(g_{n,K-1} + g_{n,K}\right) + f_{n+1,1} \ g_{n,K-2} + f_{n+1,0} \ g_{n,K-3}\right] s^{K-2} + \cdots$$

$$+ \left[f_{n+1,j}\left(g_{n,K-1} + g_{n,K}\right) + \sum_{i=0}^{j-1} f_{n+1,i} \ g_{n,K-1-(j-i)}\right] s^{K-j} + \cdots$$

$$+ \left[f_{n+1,K-1} \left(g_{n,K-1} + g_{n,K} \right) + \sum_{i=0}^{K-2} f_{n+1,i} \; g_{n,K-1-(K-1-i)} \right] s$$

$$+ f_{n+1,K} \left(g_{n,K-1} + g_{n,K} \right) + \sum_{i=1}^{K-1} f_{n+1,i} \; g_{n,K-1-(K-i)}$$

+ terms with negative powers of s .

Incorporation of this result into (3.3) yields

$$G_{n+1}(s) = g_{n+1,0} + g_{n+1,1} \; s + g_{n+1,2} \; s^2 + \cdots + g_{n+1,K} \; s^K$$

where

$$\left.
\begin{aligned}
g_{n+1,K} &= f_{n+1,0} \left(g_{n,K} + g_{n,K-1} \right) \\[2mm]
g_{n+1,j} &= f_{n+1,K-j} \; g_{n,K} + g_{n,K-1} \; + \; \sum_{i=0}^{K-j-1} f_{n+1,i} \; g_{n,i+j-1} \; , \\
& \hspace{4cm} j = 1,2,3,\ldots,K-1 \\[2mm]
g_{n+1,0} &= 1 - \sum_{j=1}^{K} g_{n+1,j} \; .
\end{aligned}
\right\}
\qquad (3.4)$$

A similar derivation for the M/G/1,K queue is also possible. When the waiting room has unlimited capacity, i.e. for the GI/M/1 queue, (3.1) is replaced by

$$Q_{n+1} = \left(Q_n + 1 - Y_{n+1} \right)^+ , \quad n = 1,2,3,\ldots \; .$$

If the probability generating functions of Q_n and Y_{n+1} are now defined respectively as $G_n(s)$ and $F_{n+1}(s)$, with

$$G_n(s) = \sum_{i=0}^{\infty} g_{n,i} \; s^i \; , \qquad F_{n+1}(s) = \sum_{i=0}^{\infty} f_{n+1,i} \; s^i \; ,$$

$\text{Prob}[Q_n = i] = g_{n,i}$, $i = 0,1,2,\ldots$ and $\text{Prob}[Y_{n+1} = i] = f_{n+1,i}$, $i = 0,1,2,\ldots$ then one obtains

$$G_{n+1}(s) = \Pi\left[sG_n(s)F_{n+1}(s^{-1})\right] , \quad n = 1,2,3,\ldots .$$

From this recurrence relation one obtains

$$g_{n+1,j} = \sum_{i=j-1}^{\infty} g_{n,i}\, f_{n+1,i-j+1} , \quad j = 1,2,3,\ldots$$

and

$$g_{n+1,0} = 1 - \sum_{j=1}^{\infty} g_{n+1,j} , \quad n = 1,2,3,\ldots .$$

$$\left.\begin{array}{}\\ \\ \\ \\ \\ \\ \end{array}\right\} \quad (3.4A)$$

Transient state probabilities for ergodic infinite waiting space queuing systems may be in principle found from (3.4A) in a recursive manner. However, it would be worthwhile to mention here two relevant aspects of the underlying problem. The infinite summations in (3.4A), for the purpose of machine evaluation, would be treated as truncated (i.e. finite) summations. This might be, for instance, done by truncating each summation at a point when the subsequent terms become numerically insignificant in comparison with a preassigned error bound. A rigorous error analysis for this procedure is possible. Nevertheless, a uniform bounding on the length of the truncated series is difficult to ascertain; in general, this length will vary with the number of completed recursions.

The second aspect of the problem pertains to the use of infinite waiting room models to represent real life queuing systems. Such models are often useful for their analytic simplicity and hence employed widely in theoretical studies. Physical considerations, nevertheless, suggest that for most practical queuing situations a model with a sufficiently large, but finite, waiting room would be satisfactory. Under this representation, the physical system itself will often suggest a reasonable value to be incorporated in the model as the maximum capacity of the waiting room. Neuts [1971] and Kerridge [1966], among others, have used this approach to model real life waiting line systems. Algorithm (3.4) would therefore be useful to examine a reasonable number of problems that were to be studied by (3.4A). The actual computations could follow the suggestions of Seneta ([1968], page 470). For systems in which the number of customers would appear to grow indefinitely, results from heavy traffic theory would be useful.

Consider the M/M/1,K queue, as a special case of the GI/M/1,K queue. Let the interarrival time associated with the nth arrival be exponentially distributed, with a mean $1/\lambda_n$. Also, let the mean service rate during this interarrival period be μ_n . Then

$$E\left(s^{Y_{n+1}}\right) = \frac{\lambda_{n+1}}{\lambda_{n+1} + \mu_{n+1}(1-s)} \quad .$$

Hence

$$F_{n+1}(s) = \frac{\lambda_{n+1}}{\lambda_{n+1} + \mu_{n+1}} s^K \left\{ 1 + \frac{\mu_{n+1}}{\lambda_{n+1} + \mu_{n+1}} \cdot \frac{1}{s} + \left(\frac{\mu_{n+1}}{\lambda_{n+1} + \mu_{n+1}}\right)^2 \frac{1}{s^2} + \cdots \right\} \quad . \quad (3.5)$$

Incorporation of (3.5) into (3.4) enables a direct evaluation of the time-dependent queue size distributions to be made when λ_n and μ_n are arbitrarily specified functions of n , the temporal parameter. The exactness of the method may be demonstrated by means of a specific queuing problem with $\lambda_n = \lambda$ and $\mu_n = \mu$, one for which analytic results are available (Wagner [1969], page 859).

TABLE 3.1

STATIONARY PROBABILITIES FOR A LIMITED WAITING

SPACE SIMPLE QUEUE

Number of customers in the system	Stationary probability		
	Simulated	Numerical	Analytic
0	0.494280	0.500246	0.500246
1	0.253120	0.250122	0.250122
2	0.125800	0.125060	0.125060
3	0.063280	0.062530	0.062530
4	0.031960	0.031265	0.031265
5	0.015560	0.015633	0.015633
6	0.008320	0.007816	0.007816
7	0.004080	0.003908	0.003908
8	0.002040	0.001954	0.001954
9	0.001080	0.000977	0.000977
10	0.000480	0.000489	0.000489

Table 3.1 displays the stationary results for a simple queue with $\lambda = 1$, $\mu = 2$ and $K = 10$. A consideration of the numerical errors appears in Appendix B. Estimates of stationary state probabilities obtained by simulating the queue, the numerically evaluated probabilities based on (3.4) and the analytically computed results are listed. An IBM 370/165 system took 1 minute and 23.4 seconds to simulate the queue, to produce only reasonable estimates of the state probabilities. The GPSS 360 model employed for this purpose appears in Appendix A. Com-

putations based on (3.4) were accomplished in 0.16 seconds. A FORTRAN
IV listing of the associated program is given also in Appendix A.

3.4   A BULK QUEUE WITH LIMITED WAITING ROOM

The bulk queuing system $GI^X/M^Y/1,K$ is modeled by the finite
Markov chain (2.5). The system has general batch input and a single
server who services random sized groups with exponentially distributed
service times.   From

$$Q_{n+1} = \left[K - Y_{n+1} + (Q_n + X_n - K)^- \right]^+ , \qquad n = 1,2,3,\ldots$$

and $Q_1 = z$, $0 \leq z \leq K$ it follows that

$$g_{n+1} = \Pi \left(f_{n+1} \ominus (g_n \ h_n) \right) , \qquad n = 1,2,3,\ldots \qquad (2.8)$$

with g_n , f_{n+1} and h_n denoting the distributions of Q_n ,
$K - Y_{n+1}$ and $X_n - K$ respectively, and with $g_1 = \delta_z$; g_n , f_{n+1} ,
$h_n \in M$. Mapping the elements in (2.8) S-homomorphically one obtains

$$G_{n+1}(s) = \Pi \left[F_{n+1}(s) \ominus (G_n(s) \ H_n(s)) \right] , \qquad n = 1,2,3,\ldots \qquad (3.6)$$

with $G_1(s) = s^z$. Now,

$$F_{n+1}(s) = E \left(s^{K-Y_{n+1}} \right)$$

$$= s^K \ \psi \left(\mu_{n+1} - \mu_{n+1} \ c_{n+1}(1/s) \right) , \qquad |s| \leq 1$$

where $\psi(\cdot)$ is the Laplace-Stieltjes transform of the interarrival time distribution, and $c_{n+1}(s)$ the current p.g.f. of the capacity of the server. Let the p.g.f. of X_n be denoted by $X_n(s) = \sum_{i=0}^{\infty} d_{n,i} \, s^i$. Also let

$$F_{n+1}(s) = f_{n+1,0} \, s^K + f_{n+1,1} \, s^{K-1} + f_{n+1,2} \, s^{K-2} + \cdots$$

$$\cdots + f_{n+1,K} + f_{n+1,K+1} \, s^{-1} + \cdots .$$

Hence, if $G_n(s)$ equals $\sum_{i=0}^{K} g_{n,i} \, s^i$, then

$$G_n(s) \, H_n(s) = G_n(s) \, s^{-K} \, X_n(s)$$

$$= g_{n,0} \, d_{n,0} \, s^{-K} + \left(g_{n,0} \, d_{n,1} + g_{n,1} \, d_{n,0}\right) s^{-K+1}$$

$$+ \left(g_{n,0} \, d_{n,2} + g_{n,1} \, d_{n,1} + g_{n,2} \, d_{n,0}\right) s^{-K+2} + \cdots$$

$$+ \left(g_{n,0} \, d_{n,j} + g_{n,1} \, d_{n,j-1} + \cdots + g_{n,j} \, d_{n,0}\right) s^{-K+j} + \cdots$$

$$+ \left(g_{n,0} \, d_{n,K-1} + g_{n,1} \, d_{n,K-2} + \cdots + g_{n,K-1} \, d_{n,0}\right) s^{-1}$$

$$+ \text{ terms with non-negative powers of } s .$$

Therefore,

$$\Theta\left(G_n(s) \, H_n(s)\right) = g_{n,0} \, d_{n,0} \, s^{-K} + \left(g_{n,0} \, d_{n,1} + g_{n,1} \, d_{n,0}\right) s^{-K+1}$$

$$+ \left(g_{n,0} \, d_{n,2} + g_{n,1} \, d_{n,1} + g_{n,2} \, d_{n,0}\right) s^{-K+2} + \cdots$$

$$+ \left(g_{n,0} \, d_{n,j} + g_{n,1} \, d_{n,j-1} + \cdots + g_{n,j} \, d_{n,0}\right) s^{-K+j} + \cdots$$

$$+ \left(g_{n,0}\, d_{n,K-1} + g_{n,1}\, d_{n,K-2} + \cdots + g_{n,K-1}\, d_{n,0}\right) s^{-1}$$

$$+ 1 - \sum \text{ (all previous coefficients)}.$$

Let $\theta\!\left(G_n(s)\, H_n(s)\right)$ be represented by

$$a_{n,0}\, s^{-K} + a_{n,1}\, s^{-K+1} + a_{n,2}\, s^{-K+2} + \cdots + a_{n,K-1}\, s^{-1} + a_K \, ,$$

where $a_{n,j} = \sum\limits_{i=0}^{j} g_{n,i}\, d_{n,j-i} \, , \quad j = 0,1,2,\ldots,K-1$

and $a_{n,K} = 1 - \sum\limits_{j=0}^{K-1} a_{n,j} \, .$

Now,

$$F_{n+1}(s)\, \theta\!\left(G_n(s)\, H_n(s)\right) = f_{n+1,0}\, a_{n,K}\, s^{K} + \left(f_{n+1,0}\, a_{n,K-1} + f_{n+1,1}\, a_{n,K}\right) s^{K-1}$$

$$+ \left(f_{n+1,0}\, a_{n,K-2} + f_{n+1,1}\, a_{n,K-1} + f_{n+1,2}\, a_{n,K}\right) s^{K-2} + \cdots$$

$$+ \left(f_{n+1,0}\, a_{n,K-j} + f_{n+1,1}\, a_{n,K-j+1} + \cdots + f_{n+1,j}\, a_{n,K}\right) s^{K-j} + \cdots$$

$$+ \left(f_{n+1,0}\, a_{n,0} + f_{n+1,1}\, a_{n,1} + \cdots + f_{n+1,K-1}\, a_{n,K-1} + f_{n+1,K}\, a_{n,K}\right)$$

$$+ \text{ terms with negative powers of } s \, .$$

Hence, $G_{n+1}(s) = \Pi\!\left(F_{n+1}(s)\, \theta\!\left(G_n(s)\, H_n(s)\right)\right)$

$$= \sum\limits_{i=0}^{K} g_{n+1,i}\, s^{i}$$

where

$$g_{n+1,K} = a_{n,K}\, f_{n+1,0}$$

$$g_{n+1,K-1} = a_{n,K}\, f_{n+1,1} + a_{n,K-1}\, f_{n+1,0}$$

$$g_{n+1,K-2} = a_{n,K}\, f_{n+1,2} + a_{n,K-1}\, f_{n+1,1} + a_{n,K-2}\, f_{n+1,0}$$

$$\vdots$$

$$g_{n+1,K-j} = a_{n,K}\, f_{n+1,j} + a_{n,K-1}\, f_{n+1,j-1} + \cdots + a_{n,K-j}\, f_{n+1,0}$$

$$\vdots$$

$$g_{n+1,0} = 1 - \sum_{i=1}^{K} g_{n+1,i}$$

$$(3.7)$$

with

$$a_{n,0} = g_{n,0}\, d_{n,0}$$

$$a_{n,1} = g_{n,0}\, d_{n,1} + g_{n,1}\, d_{n,0}$$

$$a_{n,2} = g_{n,0}\, d_{n,2} + g_{n,1}\, d_{n,1} + g_{n,2}\, d_{n,0}$$

$$\vdots$$

$$a_{n,j} = g_{n,0}\, d_{n,j} + g_{n,1}\, d_{n,j-1} + \cdots + g_{n,j}\, d_{n,0}$$

$$\vdots$$

$$a_{n,K-1} = g_{n,0}\, d_{n,K-1} + g_{n,1}\, d_{n,K-2} + \cdots + g_{n,K-1}\, d_{n,0}$$

$$a_{n,K} = 1 - \sum_{i=0}^{K-1} a_{n,i} \; .$$

Specifically, suppose that a bulk queue with compound Poisson input is being considered. Let the distribution of the size of the nth arriving batch of customers be geometric, such that

$$\text{Prob[batch size} = \nu] = (1-p_n)p_n^{\nu-1} , \quad \nu = 1,2,3,\ldots, \quad 0 < p_n < 1 .$$

Let the interarrival time of this batch be distributed as $1 - \exp(-\lambda_n t)$ $(t > 0)$. Let the service rate during this interarrival period be μ_n, and the associated service capacity be distributed geometrically as

$$\text{Prob[capacity} = \nu] = (1 - q_n)q_n^{\nu-1} , \quad \nu = 1,2,3,\ldots, \quad 0 < q_n < 1 .$$

Hence,

$$F_{n+1}(s) = \frac{\lambda_{n+1} s^K}{\lambda_{n+1} + \mu_{n+1}} + \frac{\mu_{n+1} \lambda_{n+1}(1 - q_{n+1})s^{K-1}}{(\lambda_{n+1} + \mu_{n+1})^2}$$

$$+ \frac{\mu_{n+1} \lambda_{n+1}(1 - q_{n+1})(\lambda_{n+1} q_{n+1} + \mu_{n+1})s^{K-2}}{(\lambda_{n+1} + \mu_{n+1})^3} + \cdots$$

$$+ \frac{\mu_{n+1} \lambda_{n+1}(1 - q_{n+1})(\lambda_{n+1} q_{n+1} + \mu_{n+1})^{j-1} s^{K-j}}{(\lambda_{n+1} + \mu_{n+1})^{j+1}} + \cdots .$$

Also,

$$H_n(s) = (1 - p_n) \Big/ \Big[(1 - p_n s)s^{K-1}\Big] .$$

Any non-stationarities may now be readily modeled by the parameters $\{\lambda_n, \mu_n, p_n, q_n\}$. Incorporation of these results into (3.7) yields the time-dependent probability distributions for the number of

TABLE 3.2

THE GROWTH OF QUEUE SIZE[*] IN A BULK QUEUE

n	expected queue size	n	expected queue size
1	0 (empty system)	26	2.262441
2	0.855369	27	2.262448
3	1.353801	28	2.262455
4	1.674696	29	2.262460
5	1.885333	30	2.262461
6	2.022345	31	2.262463
7	2.110283	32	2.262464
8	2.166222	33	2.262465
9	2.201650	34	2.262465
10	2.224048	35	2.262465
11	2.238199	36	2.262465
12	2.247137	37	2.262465
13	2.252784	38	2.262465
14	2.256351	39	2.262465
15	2.258603	40	2.262465
16	2.260024	41	2.262465
17	2.260925	42	2.262465
18	2.261493	43	2.262465
19	2.261850	44	2.262465
20	2.262077	45	2.262465
21	2.262218	46	2.262465
22	2.262310	47	2.262465
23	2.262367	48	2.262465
24	2.262404	49	2.262465
25	2.262426	50	2.262465

---

[*] Including those in service

customers in the system.

Consider now a particular problem which, for simplicity, is
is assumed to involve only time-independent parameters. A queue is
formed by a Poisson stream of batch arrivals, the mean rate of

arrival, λ , being one batch per unit time. Service time is distributed exponentially, the mean $(1/\mu)$ being 0.5 time units. Let the number of customers arriving in a batch be distributed geometrically, with parameter $p = 0.5$. Similarly, let the capacity of the server to accept a random number of customers as a batch have a geometric distribution, with $q = 0.5$. Also, let the whole system accommodate a maximum of ten customers.

Table 3.2 shows the growth of the queuing process in the above system, when we start with an empty system. The results are given in terms of the expected number of customers present in the system immediately before the arrival of the nth batch, $n = 1,2,...,50$. The numbers listed were computed from state probabilities, generated numerically in a recursive manner from (3.7). The execution of this process took 0.20 seconds on an IBM 370/165 system. The FORTRAN IV program used for these calculations is listed in Appendix A. A simulation of the same queuing system which gave only approximate estimates for the stationary state probabilities for this queue took 1 minute and 34.8 seconds to simulate 25000 group arrivals. The GPSS 360 simulator employed for this purpose appears in Appendix A.

In the above example, the time to compute the state probabilities $\{g_{n,j}\}$ would increase with increase in the size of the waiting room, as one would expect. However, for problems allowing up to a few hundred customers in the waiting room, this time remains reasonably small. Typical computational times are shown in Table 3.3. Listed are several waiting room sizes (K) and the corresponding times to execute one hundred recursions of (3.7) on an IBM 370/165 system.

The time-dependent behaviour of the *finite* GI/M/1 (or M/G/1) queue may be computed in principle by evaluating the matrix-convolution products of the underlying Markov transition matrices.

TABLE 3.3

WAITING ROOM SIZE VS. EXECUTION TIME

Waiting room size K	Execution time in seconds
10	0.28
20	0.50
40	1.20
80	4.07
160	15.41
320	110.70

Such matrix-convolution formalisms, to handle non-stationary Markov chains, have been expounded by Harary, Lipstein and Styan [1970] and by Bithell [1971]. It may be noted, however, that if a queue with a waiting room of size K is to be studied, as many as $(K+1)^2$ convolution product evaluations would be required at every step. The approach of employing the probability sweeping operators, which eventually lead to algorithms resembling (3.4), would often reduce total computations considerably in this respect. The savings result primarily from the elimination of operations involving matrix elements that are identically equal to zero. For a problem incorporating triangular Markov transition matrices, for example, this reduction would be approximately 50%. The savings would be significant particularly in queuing problems with large waiting rooms.

3.5   QUEUES WITH VARIABLE ARRIVAL RATE

Among the various waiting line situations, systems with variable
input conditions are of substantial practical significance.   Clarke
[1953] and Luchak [1956] have analytically studied simple queues with
exponentially distributed interarrival times, and having a mean
arrival rate which varies with time.   Leese [1964] points out certain
computational difficulties inherent in the application of these results to actual problems.   Quoting from Leese:

> "Two difficulties arise in applying Clarke's method
> to solve a specific numerical problem.   In the first
> place, the kernel of Clarke's integral equation tends to
> infinity at one end of the range of integration.   Secondly,
> Clarke's formulae contain expressions of the form

$$Q_n(t) = P_n(t) \exp\left[t + \int_0^t \rho(s)\,ds \right] . \tag{1}$$

> ... exponential expression in (1) may have a value as great
> as e^{600} .   Both these features of Clarke's solution make
> numerical solution by standard methods extremely awkward.
>
> Luchak's solution consists essentially of a Taylor's
> series expansion of $Q_n(t)$ about $t = 0$.   Because Q_n
> increases so rapidly with t ,   an extremely large number
> of terms of Taylor's series would be required to obtain
> satisfactory answers for values of t as high as 400 ."

The numerical method suggested by Leese [1964] to attack the
variable arrival rate problem seems more efficient than the use of
Monte Carlo simulation.   It involves, nevertheless, an approximate
numerical solution of a certain integral equation, and is applicable
only to the rather special case of a simple Markovian queue.   The
problem may be tackled more profitably by employing the notions set
forth earlier in this chapter.   This point will be illustrated by
considering a specific example.

We begin by recalling that sets of equations generated under the
proposed approach, such as (3.7), allow for the incorporation of

arbitrary functional forms involving the temporal parameter, n . This is accomplished by means of the parameters λ_n , μ_n , p_n , q_n , etc., as they appear in these equations. Suppose now we consider a simple queue with a Poisson input, with a cyclically varying (periodic) mean arrival rate λ_n . Let service times be exponential, with a constant mean, $1/\mu$. Also, suppose that the system as a whole is designed to permit at most K customers to be present at any time.

Let the successive epochs of arrivals be numbered as $n = 1, 2, 3, \ldots$, where $n = 1$ is the instant that the queue begins to operate. Let the periodicity in the arrival pattern be modeled by

$$\lambda_n = 2.0 + 1.5 \sin(2\pi n/T) ,$$

where T is the period of the cycle, expressed as an integer. Also, let μ be equal to 2.0 , and let K be 10 . The progress of the queuing process may now be studied directly, utilizing (3.4).

Table 3.4 shows the numerically evaluated results for this system. A periodicity in λ_n appears to induce a similar periodicity in the expected queue size, an effect that would perhaps be anticipated from intuitive grounds. Figure 3.1 displays these results graphically.

A variety of effects associated with periodically changing arrival rates can be studied using (3.4). An interesting effect was noted when we let the period T change, from queue to queue, while the values of the other parameters remained unaltered. The observed amplitude of the induced periodic oscillations in the mean queue size was a direct function of T . Figure 3.2 exhibits the results. While this phenomenon is not surprising, we have not seen it explicitly mentioned in the literature. It may be naively remarked, on the basis of Figure 3.2, that a waiting line with limited waiting room is

39

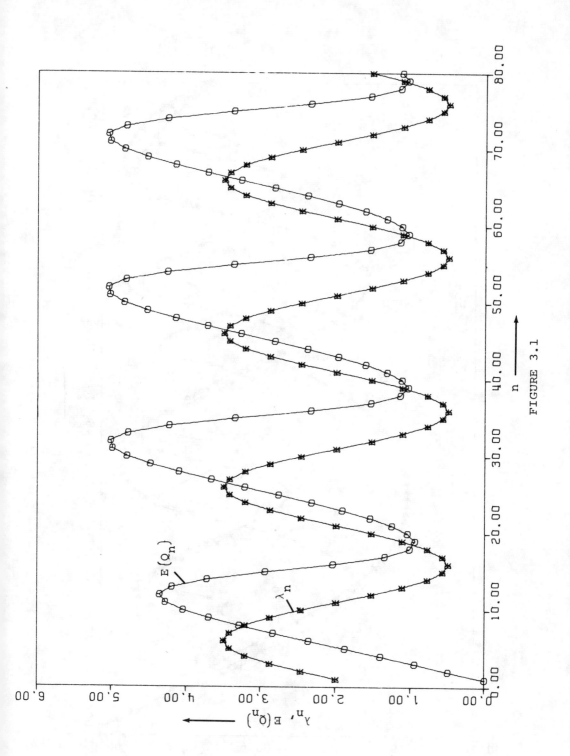

FIGURE 3.1

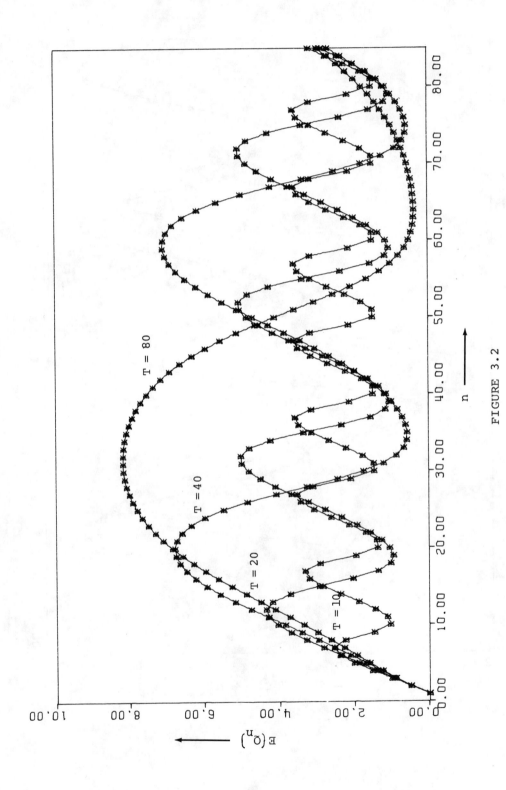

FIGURE 3.2

TABLE 3.4

VARIATIONS IN EXPECTED QUEUE SIZE WITH PERIODIC INPUT

n	λ_n	$E[Q_n]$	n	λ_n	$E[Q_n]$
1	2.463525	0.000000	31	1.536486	4.998431
2	2.881677	0.551924	32	1.118332	4.772689
3	3.213524	1.049587	33	0.786483	4.223559
4	3.426583	1.541062	34	0.573420	3.342372
5	3.499999	2.031623	35	0.500001	2.326999
6	3.426585	2.513680	36	0.573411	1.540964
7	3.213528	2.973295	37	0.786466	1.145211
8	2.881680	3.391089	38	1.118309	1.044744
9	2.463529	3.740674	39	1.536460	1.126928
10	2.000003	3.985364	40	1.999985	1.329905
11	1.536479	4.073775	41	2.463507	1.620910
12	1.118325	3.932946	42	2.881663	1.978284
13	0.786479	3.492571	43	3.213505	2.384008
14	0.573418	2.755833	44	3.426574	2.821189
15	0.500001	1.910928	45	3.499999	3..273092
16	0.573414	1.270810	46	3.426594	3.722391
17	0.786470	0.963926	47	3.213542	4.149907
18	1.118316	0.909147	48	2.881700	4.532434
19	1.536467	1.014537	49	2.463550	4.839464
20	1.999992	1.229967	50	2.000024	5.029365
21	2.463516	1.528281	51	1.536513	5.045528
22	2.881670	1.890515	52	1.118351	4.814566
23	3.213519	2.300052	53	0.786492	4.259712
24	3.426581	2.740749	54	0.573430	3.371388
25	3.499999	3.196316	55	0.500001	2.347682
26	3.426587	3.649674	56	0.573405	1.554447
27	3.213532	4.081747	57	0.786445	1.154280
28	2.881687	4.469301	58	1.118284	1.051524
29	2.463536	4.781670	59	1.536436	1.132545
30	2.000010	4.976980	60	1.999966	1.334892

analogous to a simple electrical circuit, connected to a current source, as portrayed in Figure 3.3. It is possible that the construction of models utilizing analogs of this nature might be useful and instructive in the study of waiting lines and their networks.

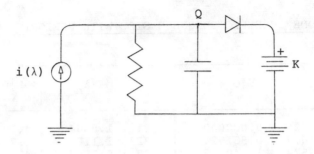

FIGURE 3.3   THE LIMITED WAITING ROOM QUEUE ANALOG

3.6   A HEURISTIC EXPERIMENT

It seems natural to ask whether the wide fluctuations in the expected queue size, as encountered in the example above, can be reduced by strategically manipulating the service rate.   This is something that would seem to be rather hard to answer by analytic approaches.   Using the numerical approach to experiment with a simple queuing problem having periodic arrivals and service, it was found that this question has a relatively simple answer, and one that is intuitively plausible.

The behaviour of the simple Markovian queue characterized in the previous section was examined, now subject to a periodically changing mean service rate μ_n .   The mean service rate was modeled by

$$\mu_n = 2.0 + 1.5 \sin\left(2\pi(n+\theta)/T\right) ,$$

where θ specified the phase difference between the input and service rates.   The state probabilities for this queue were evaluated numeri-

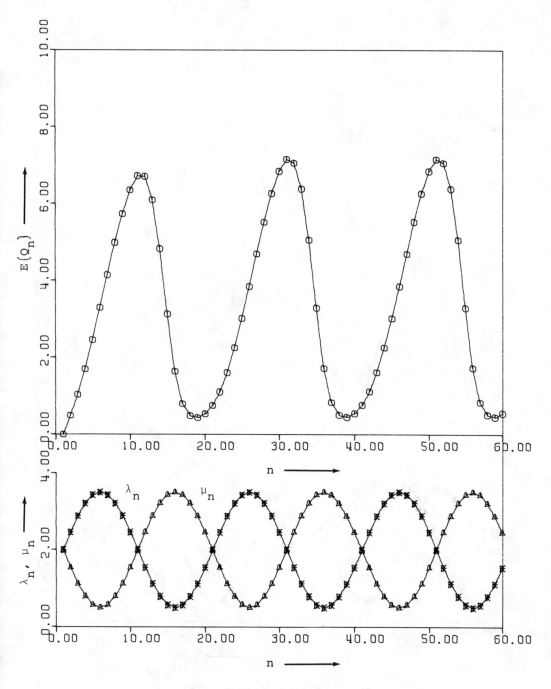

FIGURE 3.4   θ = T/2

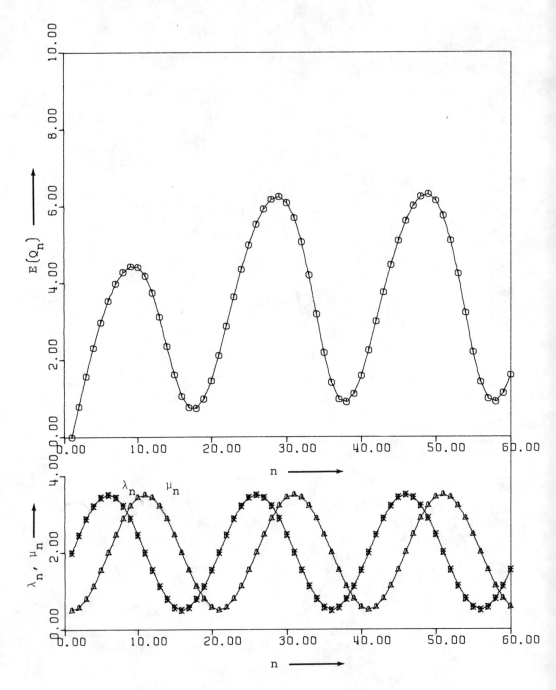

FIGURE 3.5 $\theta = T/4$

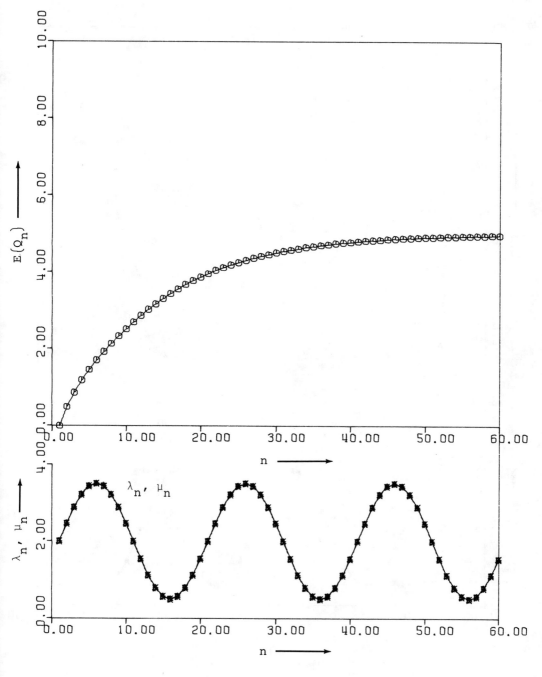

FIGURE 3.6 θ = 0

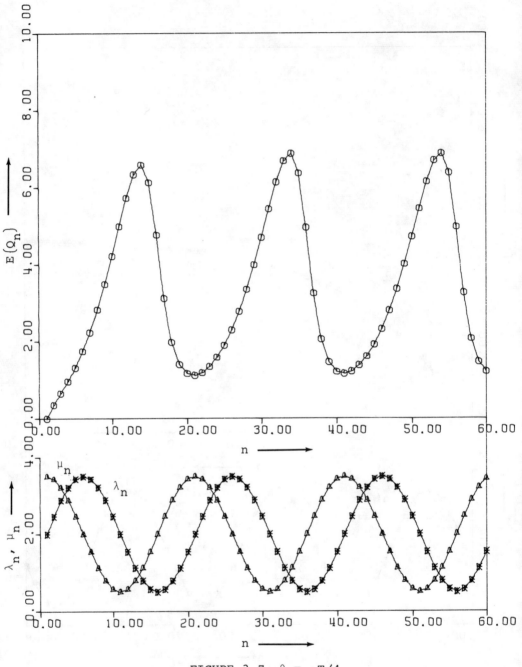

FIGURE 3.7   θ = -T/4

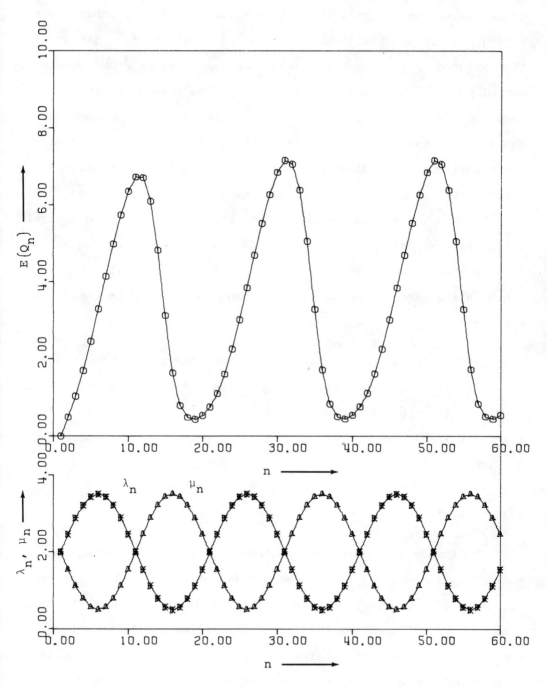

FIGURE 3.8 $\theta = -T/2$

cally. Figures 3.4 - 8 display the computational results. From these graphs it may be observed that with cyclically varying input and servicing rates, each having the same period T , the fluctuations in the expected queue size are maximum when λ_n and μ_n are completely out of phase. The fluctuations, save for the effect of transients, become minimum when λ_n and μ_n are in phase. This phenomenon was characteristically observed, also for "square wave" input and servicing conditions, regardless of the relative amplitudes of the oscillations in the arrival and servicing rates. It may be inferred from this that in congestion phenomena with periodically varying arrival rates, such as daily street traffic fluctuations, the servicing intensity should be set in phase with the arrival intensity if fluctuations in the expected volume due to congestion are to be kept to a minimum.

If the time dependence of the arrival rate is not accurately known in advance, some delay is to be expected in increasing the service rate to match an increase in the arrival rate. In this case, it may be desirable to control service rate by using a state-dependent service rate.

3.7   OPERATIONAL SOLUTIONS VS. EXACT RESULTS

The stride of queuing theory toward increasingly formalized mathematical models is sometimes justifiably questioned by the practitioners of operational research. Lee [1966] in his book "Applied Queueing Theory" promotes the use of simplified models to obtain operational solutions to real-world problems, as opposed to becoming submerged in an exact and involved analysis. An *operational* solution is one which need not be exact, but must not be so inaccurate as to lead to wrong decisions in the practical problem at hand. Lee argues:

"... existing queueing theory is useful. In spite of its
mathematical complexities on the one hand and its practical inadequacies on the other, the theory in its present
form can provide operational solutions in many situations.
This requires however that one accepts the necessity of
approximations ... by the use of oversimplified models.
Another way of saying this is that the first type of
approximation is to obtain operational solutions by applying models which are, a priori, the wrong ones but which
are simple to deal with."

Along these lines a reasonable, though perhaps naive, approach to
the study of a bulk queuing situation would be to avoid the complexities arising out of randomness in the size of the arriving batches,
and in the capacities that become available during the servicing
cycles. One may choose to replace the chance variables by appropriately "lumping" their values. For instance, it may be assumed
that at each arrival epoch, a batch of fixed size arrives; the size is
equal to the mean size of the original batches. Likewise, the random
valued capacities may be replaced by a fixed number, equal to the
corresponding value of the expected capacity. Rather than use the
mean capacity, one might do better to find some other fixed capacity
other than the mean capacity, which would be a "certainty equivalent"
in the sense of Charnes and Cooper [1959].

Table 3.5 presents the numerically evaluated stationary results,
as a function of the traffic intensity, for a typical case from a class
of bulk queues with limited waiting space studied. The original distributions of incoming batch sizes and the available service capacities
were assumed to be geometric, each with a parameter 0.5 . Both interarrival and service times were assumed to be distributed exponentially,
with the mean rate for service being unity. The maximum number of
customers allowed in the system was ten. From the table the effect of
this approximate modelling, using the above lumping procedure, is
apparent. For the class studied it is inferred that the lumping procedure yields answers that are underestimates when the traffic

TABLE 3.5

OPERATIONAL vs. EXACT RESULTS FOR A BULK QUEUE

Traffic intensity	Expected (stationary) queue size	
	the original queue	the lumped queue
0.5	2.262465	1.809618
1.0	4.999986	4.999990
1.5	6.764890	7.154872
2.0	7.737515	8.190463
2.5	8.299560	8.716008

intensity is less than one, and overestimates when it is greater than one.

By making comparisons of the above nature, conclusions might be drawn relatively easily about the effectiveness of such approximating techniques in a given problem. It should be noted that this can be done in advance of gathering specific data on the distributions or other features involved in the problem by performing the comparison using typical, rather than actual, data. In making this comparison, the exact results can be generated numerically in a reasonably large number of situations.

CHAPTER IV

CONCLUSIONS AND DIRECTIONS FOR FURTHER RESEARCH

The central feature of the present work is an algorithm for computing numerically the time dependent probabilities for discrete parameter Markov chains. The algorithm has been obtained by employing the concept of sweeping of probabilities in certain subspaces of the state space. It has been applied to examine the behaviour of a sizable class of bulk queues and queues with non-stationary input and service conditions.

The present method is mathematically equivalent to the matrix-convolution product method, but it is, in many cases, computationally more efficient than the matrix method. The rate of growth of the overall effort in sample problems considered suggests that the present method will be applicable to examine moderately large sized problems. Like all numerical methods, the present method would suffer from computational errors. Estimates on the upper bounds on errors committed in the recursions carried out may be obtained (see Appendix B), though these may sometimes be pessimistic. Perhaps a more meaningful approach to the examination of errors would be to adopt the statistical approach to error analysis developed by Henrici [1962]. It is felt, neverthe-less, that the use of double precision should permit generation of satisfactory results for all reasonably sized problems.

Methods for obtaining upper and lower bounds on the expected queue size for bulk queues are considered in Appendix C as extensions of some available analytic inequalities. It is noted that reasonably sharp upper bounds are easy to evaluate numerically. Scope exists, nevertheless, for further work on the efficient evaluation of the lower bounds. Two specific problems are proposed in this connection.

The limitation to the scope of the present work lies in the availability of suitably imbedded Markov chains. Further studies may be undertaken to attempt to imbed a wider and richer variety of congestion problems. In particular, this should be done for multiple server bulk queues. Such attempts will be greatly facilitated by an improved understanding of point processes. Can the notion of sweeping of probabilities be extended, both in the analytic and the computational sense, to study semi-Markov processes? This may well be possible in the discrete parameter situation.

REFERENCES

1. Bailey, N.T.J. [1954], "On Queueing Processes with Bulk Service", J. Roy. Statist. Soc. Ser. B, 16, pp. 80-87.

2. Bailey, N.T.J., The Elements of Stochastic Processes with Applications to the Natural Sciences, Wiley, New York, 1964.

3. Baxter, G. [1960], "An Analytic Problem Whose Solution Follows from a Simple Algebraic Identity", Pacific J. Math. 10, pp. 731-742.

4. Bhat, U.N. [1964], "Imbedded Markov Chain Analysis of Single Server Bulk Queues", J. Austral. Math. Soc. 4, pp. 244-263.

5. Bhat, U.N. [1969], "Sixty Years of Queueing Theory", Management Sci. 15, pp. B-280 to B-294.

6. Bithell, J.F. [1971], "Some Generalized Markov Chain Occupancy Processes and their Application to Hospital Admission Systems", Rev. Inst. Internat. Statist. 39, pp. 170-184.

7. Charnes, A. and Cooper, W.W. [1959], "Chance-constrained Programming", Management Sci. 6, pp. 73-80.

8. Chaudhry, M.L. and Templeton, J.G.C. [1971], "The Theory of Bulk-Arrival Bulk-Service Queues", (unpublished).

9. Clarke, A.B. [1953], "The Time Dependent Waiting Line Problem", Univ. Michigan Report M720-1R39.

10. Cohen, J.W. [1967], "On Two Integral Equations of Queueing Theory", J. Appl. Probability, 4, pp. 343-355.

11. Cohen, J.W., The Single Server Queue, North-Holland Publishing
         Company, Amsterdam, 1969.

12. Cox, D.R. [1955], "The Analysis of Non-Markovian Stochastic
         Processes by the Inclusion of Supplementary Variables,
         Proc. Cambridge Philos. Soc. 51, pp. 433-441.

13. Fox, L. and Mayers, D.F., Computing Methods for Scientists and
         Engineers, Clarendon Press, Oxford, 1968.

14. Ghosal, A., Some Aspects of Queueing and Storage Systems,
         Springer-Verlag, 1970.

15. Harary, F., Lipstein, B. and Styan, G.P.H. [1970], "A Matrix
         Approach to Nonstationary Chains", Operations Res. 18,
         pp. 1168-1181.

16. Hasofer, A.M. [1964], "On the Single-Server Queue with Non-
         homogeneous Poisson Input and General Service Time",
         J. Appl. Probability, 1, pp. 369-384.

17. Henrici, P., Lecture Notes on Elementary Numerical Analysis,
         Summer Inst. for Numerical Analysis, U.C.L.A., 1962.

18. Hillier, F.S. and Boling, R.W. [1967], "Finite Queues in Series
         with Exponential or Erlang Service Times - A Numerical
         Approach", Operations Res. 15, pp. 286-303.

19. Hirasawa, K. [1971], "Numerical Solutions of Bulk Queues via
         Imbedded Markov Chain", Electrical Eng. in Japan, 91, pp.
         127-136.

20. Iglehart, D. and Whitt, W. [1970a], "Multiple Channel Queues in
         Heavy Traffic, I", Advances in Appl. Probability 2, pp.150-177.

21. Iglehart, D. and Whitt, W. [1970b], "Multiple Channel Queues in

Heavy Traffic, II: Sequences, Networks and Batches",
Advances in Appl. Probability 2, pp. 355-369.

22. Keilson, J. [1962], "The General Bulk Queue as a Hilbert
    Problem", J. Roy. Statist. Soc. Ser. B, 24, pp. 344-358.

23. Kendall, D.G. [1951], "Some Problems in the Theory of Queues",
    J. Roy. Statist. Soc. Ser. B, 13, pp. 151-171.

24. Kerridge, D. [1966], "A Numerical Method for the Solution of
    Queueing Problems", New J. Statist. Operational Res. 2,
    pp. 3-13.

25. Kingman, J.F.C. [1962], "Spitzer's Identity and Its Use in
    Probability Theory", J. London Math. Soc. 37, pp. 309-316.

26. Kingman, J.F.C. [1962a], "Some Inequalities for the Queue
    GI/G/1", Biometrika, 49, pp. 315-324.

27. Kingman, J.F.C. [1965], "The Heavy Traffic Approximation
    in the Theory of Queues", Proc. Sym. on Congestion Theory,
    eds. Smith, W.L. & Wilkinson, W.E., Univ. of North Carolina
    Press, Chapel Hill, N.C.

28. Kingman, J.F.C. [1966], "On the Algebra of Queues", J. Appl.
    Probability, 3, pp. 285-326.

29. Kingman, J.F.C. [1970], "Inequalities in the Theory of Queues",
    J. Roy. Statist. Soc. Ser. B, 32, pp. 102-110.

30. Kinney, J.R. [1962], "A Transient Discrete Time Queue with
    Finite Storage", Ann. Math. Statist. 33, pp. 130-136.

31. Kleinrock, L., Muntz, R.R. and Hsu, J. [1971], "Tight Bounds on
    the Average Response Time for Time-shared Computer Systems",
    Proc. of the IFIP Congress 71, Ljubljana, Yugoslavia.

32. Klimko, E.M. and Neuts, M.F. [1972], "The Single Server Queue
    in Discrete Time-Numerical Analysis II", Purdue Mimeo Series
    No. 276, Dept. of Statist., Purdue Univ., W. Lafayette, In.

33. Lee, A.M., Applied Queueing Theory, St. Martin's Press,
    (Macmillan), 1966.

34. Leese, E.L. [1964], "Numerical Methods of Determining the
    Transient Behaviour of Queues with Variable Arrival Rate",
    Queuing Theory, The English Universities Press Ltd., London.

35. Leese, E.L. and Boyd, D.W. [1966], "Numerical Methods of Deter-
    mining the Transient Behaviour of Queues with Variable
    Arrival Rates", J. Canad. Operations Res. Soc., 4, pp. 1-13.

36. Lindley, D.V. [1952], "The Theory of Queues with a Single
    Server", Proc. Cambridge Philos. Soc. 48, pp. 277-289.

37. Luchak, G. [1956], "The Solution of the Single Channel Queueing
    Equations Characterized by a Time-dependent Poisson-
    distributed Arrival Rate and a General Class of Holding
    Times", Operations Res., 4, pp. 711-732.

38. Mack, C. [1967], "The Pitfall in Assuming Exponential Service
    Time in Queues", New J. Statist. Operational Res., 3,
    pp. 17-21.

39. Marshall, K.T. [1968], "Some Inequalities in Queueing", Opera-
    tions Res., 16, pp. 651-665.

40. Miller, R.G. [1959], "A Contribution to the Theory of Bulk
    Queues", J. Roy. Statist. Soc. Ser. B, 21, pp. 320-337.

41. Muskhelishvili, N.I., Singular Integral Equations, Erven P.
    Noordhoff, NV Groningen, Netherlands, 1953.

42. Newell, G.F., Applications of Queueing Theory, Chapman and Hall, London, 1971.

43. Neuts, M.F. [1971], "The Single Server Queue in Discrete Time-Numerical Analysis I", Purdue Mimeo Series No. 270, Dept. of Statist., Purdue Univ., W. Lafayette, In.

44. Page, E.S. [1965], "Computers and Congestion Problems", Proc. Sym. on Congestion Theory, eds. Smith, W.L. & Wilkinson, W.E., Univ. of North Carolina Press, Chapel Hill, N.C.

45. Pollaczek, F. [1957], "Problèmes stochastiques posés par le phénomeñe de formation d'une queue d'attente à un guichet et par des phénomènes apparentés. Mémor. Sci. Math. 136. Gauthier-Villars, Paris, 1957.

46. Roes, P.B.M. [1970], "The Finite Dam", J. Appl. Probability, 7, pp. 316-326.

47. Saaty, T.L., Elements of Queueing Theory, McGraw-Hill, New York, 1961.

48. Satyamurty, P.R. [1965], "Queuing with Balking - A Simple Method to Study the Transient Behaviour", Operations Res., 13, pp. 329-333.

49. Seneta, E. [1968], "Finite Approximations to Infinite Non-negative Matrices, II: Refinements and Applications", Proc. Cambridge Philos. Soc., 64, pp. 465-470.

50. Smith, W.L. [1953], "On the Distribution of Queueing Times", Proc. Cambridge Philos. Soc., 49, pp. 449-461.

51. Spitzer, F. [1956], "A Combinatorial Lemma and its Application to Probability Theory", Trans. Amer. Math. Soc. 82, pp. 323-339.

52. Teghem, J., Loris-Teghem, J., Lambotte, J.P., Modèles d'Attente
    M/G/1 et GI/M/1 à Arrivées et Services en Groupes,
    Springer-Verlag Berlin, Heidelberg, New York, 1969.

53. Wagner, H.M., Principles of Operations Research, Prentice-Hall,
    New Jersey, 1969.

54. Wendel, J.G. [1968], "Spitzer's Formula: A Short Proof", Proc.
    Amer. Math. Soc. 9, pp. 905-908.

APPENDIX A

PROGRAM LISTINGS

FORTRAN IV G LEVEL 18   MAIN   DATE = 71146

```
            C
            C       TIME-DEPENDENT SOLUTION OF A SIMPLE QUEUE
            C       THE TIME-DEPENDENT STATE PROBABILITIES ARE
            C       NUMERICALLY EVALUATED BY THIS PROGRAM.
            C       MAXIMUM NO. OF CALLS IN THE SYSTEM = K-1
            C       STATE K=1 CORRESPONDS TO EMPTY SYSTEM
            C
 0001               DIMENSION G(60,11),F(11)
 0002               K=11
 0003               AL=1.0
 0004               AMU=2.0
 0005               AMLM=AMU/(AL+AMU)
 0006               ALLM=AL/(AL+AMU)
 0007               DO 10 I=1,K
 0008               G(1,I)=0.0
 0009        10     CONTINUE
            C
            C       INITIALISE THE SYSTEM
            C
 0010               G(1,1)=1.0
 0011               DO 11 I=1,K
 0012               J=I-1
 0013               F(I)=ALLM*AMLM**J
 0014        11     CONTINUE
 0015               DO 15 N=1,50
 0016               G(N+1,K)=F(1)*(G(N,K)+G(N,K-1))
 0017               K2=K-2
 0018               DO 14 J=1,K2
 0019               SUM=0.0
 0020               DO 12 I=1,J
 0021               SUM=SUM+F(I)*G(N,K-J+I-2)
 0022        12     CONTINUE
 0023               G(N+1,K-J)=F(J+1)*(G(N,K)+G(N,K-1))+SUM
 0024        14     CONTINUE
 0025               S=0.0
 0026               DO 16 J=2,K
 0027               S=S+G(N+1,J)
 0028        16     CONTINUE
 0029               G(N+1,1)=1.0-S
 0030        15     CONTINUE
            C       PRINT THE STATE PROBABILITIES.
 0031               WRITE(6,200) (N,(G(N,I),I=1,11),N=1,50)
 0032        200    FORMAT(//I15,11F8.5)
 0033               STOP
 0034               END
```

```
*LOC    OPERATION  A,B,C,D,E,F,G            COMMENTS
        SIMULATE
*
*       THE   SIMPLE   QUEUE   SIMULATOR
*       THE   MAXIMUM   NO.   OF   CUSTOMERS   ALLOWED   IN   THE   SYSTEM
*       EQUALS  THE   CAPACITY   OF   STORAGE   1.
*
      1 FUNCTION    RN1,C24        EXPONENTIAL   DISTRIBUTION
0.      0     .1      .104   .2      .222   .3     .355   .4     .509   .5     .69
.6      .915  .7      1.2    .75     1.38   .8     1.6    .84    1.83   .88    2.12
.9      2.3   .92     2.52   .94     2.81   .95    2.99   .96    3.2    .97    3.5
.98     3.9   .99     4.6    .995    5.3    .998   6.2    .999   7      .9997 8
*
        GENERATE    100,FN1
        LOGIC R     5
        TABULATE    1
        TABULATE    2
        TERMINATE   1
*
        GENERATE    ,,
        SEIZE       1
        GATE LR     5
        LOGIC S     5
        SPLIT       1,VONE
        GATE LS     7
        LOGIC R     7
        ENTER       1,X1
        ENTER       2,X1
        RELEASE     1
        TERMINATE
*
 VONE   SAVEVALUE   3,V1
        SAVEVALUE   4,R1
        TEST GE     X4,X3,FILL
        SAVEVALUE   1,X3
        LOGIC S     7
        TERMINATE
 FILL   SAVEVALUE   1,X4
        LOGIC S     7
        TERMINATE
*
        GENERATE    ,,
        SEIZE       2
        GATE SNE    2
        SPLIT       1,VTWO
        GATE LS     6
        LOGIC R     6
        LEAVE       2,X2
        ADVANCE     50,FN1
        LEAVE       1,X2
        RELEASE     2
        TERMINATE
*
 VTWO   SAVEVALUE   5,V2
        SAVEVALUE   6,S2
        TEST G      X5,X6,CAP
```

```
        SAVEVALUE   2,X6
        LOGIC S     6
        TERMINATE
CAP     SAVEVALUE   2,X5
        LOGIC S     6
        TERMINATE
*
1       VARIABLE    1
2       VARIABLE    1
1       TABLE       S1,0,1,12
2       TABLE       S2,0,1,12
1       STORAGE     10
2       STORAGE     10
        START       1000,NP
        RESET
        START       25000
        END
```

63

```
FORTRAN IV G LEVEL  18                    MAIN                    DATE = 71147

          C
          C     TIME-DEPENDENT SOLUTION OF A BULK QUEUE
          C     THE TIME-DEPENDENT STATE PROBABILITIES ARE
          C     NUMERICALLY EVALUATED BY THIS PROGRAM.
          C
          C     MAXIMUM NO. OF CALLS IN THE SYSTEM # K-1
          C
0001            DIMENSION F(20),G(105,20),D(20),A(20),E(105)
0002            K=11
0003            AL=1.0
0004            AMU=2.0
0005            P=0.5
0006            Q=0.5
0007            ALPMU=AL*P+AMU
0008            SLMU=AL+AMU
0009            DO 10 I=1,K
0010            G(1,I)=0.0
0011       10   CONTINUE
          C     INITIALISE THE SYSTEM
0012            G(1,1)=1.0
0013            F(1)=AL/SLMU
0014            F(2)=AL*AMU*(1.0-P)/SLMU**2
0015            DO 11 I=3,K
0016            F(I)=F(I-1)*ALPMU/SLMU
0017       11   CONTINUE
0018            DO 18 N=1,50
0019            D(1)=0.0
0020            D(2)=1.0-Q
0021            DO 12 I=3,K
0022            D(I)=D(I-1)*Q
0023       12   CONTINUE
0024            SA=0.0
0025            K1=K-1
0026            DO 14 I=1,K1
0027            A(I)=0.0
0028            DO 15 J=1,I
0029            A(I)=A(I)+G(N,J)*D(I-J+1)
0030       15   CONTINUE
0031            SA=SA+A(I)
0032       14   CONTINUE
0033            A(K)=1.0-SA
0034            S=0.0
0035            KI=K-1
0036            DO 16 I=1,KI
0037            J=I-1
0038            G(N+1,K-J)=0.0
0039            DO 17 L=1,I
0040            IL=L-1
0041            G(N+1,K-J)=G(N+1,K-J)+F(L)*A(K-J+IL)
0042       17   CONTINUE
0043            S=S+G(N+1,K-J)
0044       16   CONTINUE
0045            G(N+1,1)=1.0-S
0046            E(N)=0.0
0047            DO 19 I=1,10
0048            AI=I
0049            E(N)=E(N)+AI*G(N,I+1)
0050       19   CONTINUE
```

FORTRAN IV G LEVEL   18   MAIN   DATE = 71147

```
0051          18    CONTINUE
              C     PRINT OUT THE STATE PROBABILITIES
0052                WRITE(6,200)    (N,(G(N,I),I=1,11),E(N),N=1,50)
0053         200    FORMAT(/I15,11F8.6,5X,F15.6)
0054                STOP
0055                END
```

```
*LOC    OPERATION  A,B,C,D,E,F,G              COMMENTS
        SIMULATE
*
*       THE  BULK  QUEUE  SIMULATOR
*       THE  MAXIMUM  NO.  OF  CUSTOMERS  ALLOWED  IN  THE  SYSTEM
*       EQUALS  THE  CAPACITY  OF  STORAGE  1.
*
     .1 FUNCTION   RN1,C24       EXPONENTIAL  DISTRIBUTION
0      0      .1    .104   .2    .222   .3    .355   .4    .509   .5    .69
.6     .915   .7    1.2    .75   1.38   .8    1.6    .84   1.83   .88   2.12
.9     2.3    .92   2.52   .94   2.81   .95   2.99   .96   3.2    .97   3.5
.98    3.9    .99   4.6    .995  5.3    .998  6.2    .999  7      .9997 8
     2 FUNCTION   RN1,D10       GEOMETRIC  DISTRIBUTION,  P=0.5
.5,1/.75,2/.875,3/.9375,4/.96875,5/.984375,6/.9921875,7/.99609375,8/
.998046875,9/1.0,10
*
        GENERATE   100,FN1
        LOGIC R    5
        TABULATE   1
        TABULATE   2
        TERMINATE  1
*
        GENERATE   ,,
        SEIZE      1
        GATE LR    5
        LOGIC S    5
        SPLIT      1,VONE
        GATE LS    7
        LOGIC R    7
        ENTER      1,X1
        ENTER      2,X1
        RELEASE    1
        TERMINATE
*
VONE    SAVEVALUE  3,V1
        SAVEVALUE  4,R1
        TEST GE    X4,X3,FILL
        SAVEVALUE  1,X3
        LOGIC S    7
        TERMINATE
FILL    SAVEVALUE  1,X4
        LOGIC S    7
        TERMINATE
*
        GENERATE   ,,
        SEIZE      2
        GATE SNE   2
        SPLIT      1,VTWO
        GATE LS    6
        LOGIC R    6
        LEAVE      2,X2
        ADVANCE    50,FN1
        LEAVE      1,X2
        RELEASE    2
        TERMINATE
*
```

```
VTWO   SAVEVALUE   5,V2
       SAVEVALUE   6,S2
       TEST G      X5,X6,CAP
       SAVEVALUE   2,X6
       LOGIC S     6
       TERMINATE
CAP    SAVEVALUE   2,X5
       LOGIC S     6
       TERMINATE
*
1      VARIABLE    FN2          RANDOMLY  SELECT  ARRIVING  BATCH  SIZE
2      VARIABLE    FN2          RANDOMLY  SELECT  SERVICE  CAPACITY
1      TABLE       S1,0,1,12
2      TABLE       S2,0,1,12
1      STORAGE     10
2      STORAGE     10
       START       1000,NP
       RESET
       START       25000
       END
```

APPENDIX B

ON THE ANALYSIS OF COMPUTATIONAL ERRORS

Use of a numerical approach to tackle a mathematical problem usually involves two aspects:

(i) the creation of the algorithm, and

(ii) an attempt to predict how this algorithm will perform.

An important element of performance here is the algorithm's stability, that is its sensitivity to perturbations, particularly those that arise from rounding errors.

For the algorithms set forth in Chapter III, it would be desirable to know the effect of such errors on the calculation of the state probabilities. In particular, for an ergodic system we would want the probabilities to converge to within reasonable limits of their steady state values before the propagation of round-off errors becomes too serious. For a system that is not ergodic it might be possible for the instability due to these propagating errors to cloud the effects caused by the system's basic probabilistic structure. In the following pages a procedure is described to study the propagation of round-off errors when utilizing algorithms such as (3.4) or (3.7). This procedure is essentially an extension of a simpler result due to Henrici [1962].

Let $\underline{g}_0$, $\underline{g}_1$, $\underline{g}_2$, ... be a sequence of vectors generated by the repeated application of the linear recurrence relation

$$\underline{g}_{n+1} = F_n[\underline{g}_n] \ , \quad n = 0,1,2,\ldots \ ,$$

where F_n is a well-defined transformation. It will suffice for the present purpose to consider the case where $F_n[g_n]$ consists of multiplying $\underline{g}_n$ by a matrix with known elements. Let the machine representations of $\underline{g}_n$ and F_n be denoted by $\hat{\underline{g}}_n$ and $\hat{F}_n$. It may therefore be seen that

$$\hat{\underline{g}}_{n+1} = \hat{F}_n[\hat{\underline{g}}_n]$$

$$= F_n[\hat{\underline{g}}_n] + \underline{\varepsilon}_n , \quad n = 0,1,2,\ldots ,$$

where $\underline{\varepsilon}_n$ denotes the local error introduced in executing the nth recursion. Analysing the arithmetic employed to compute $\underline{g}_{n+1}$ from $\underline{g}_n$ usually allows some statement to be made about $\underline{\varepsilon}_n$, as will be seen later.

Also, with the appropriate differentiability assumptions holding for $F_n[\underline{g}_n]$ and neglecting the higher order terms, one may write

$$F_n[\hat{\underline{g}}_n] - F_n[\underline{g}_n] = J_n \cdot (\hat{\underline{g}}_n - \underline{g}_n)$$

where J_n is a Jacobian defined as

$$J_n = \left\{ \begin{array}{cccc} \dfrac{\partial F_{n0}}{\partial g_{n,0}} & \dfrac{\partial F_{n0}}{\partial g_{n,1}} & \dfrac{\partial F_{n0}}{\partial g_{n,2}} \quad \cdots & \dfrac{\partial F_{n0}}{\partial g_{n,K}} \\[3ex] \dfrac{\partial F_{n1}}{\partial g_{n,0}} & \dfrac{\partial F_{n1}}{\partial g_{n,1}} & \cdots & \dfrac{\partial F_{n1}}{\partial g_{n,K}} \\[3ex] \vdots & & & \\[2ex] \dfrac{\partial F_{nK}}{\partial g_{n,0}} & \dfrac{\partial F_{nK}}{\partial g_{n,1}} & \cdots & \dfrac{\partial F_{nK}}{\partial g_{n,K}} \end{array} \right\}$$

when $\underline{g}_n = (g_{n,0} \; g_{n,1} \; \cdots \; g_{n,K})'$ and $g_{n+1,i} = F_{ni}[\underline{g}_n]$, which equals $F_{ni}[(g_{n,0} \; g_{n,1} \; g_{n,2} \; \cdots \; g_{n,K})']$, $n = 0,1,2,\ldots$.

Let the accumulated round-off error vector $\underline{r}_n$ $(n = 0,1,2,\ldots)$ be the difference between $\hat{\underline{g}}_{n+1}$ (the computed value of $\underline{g}_{n+1}$) and the true value, $\underline{g}_{n+1}$. One may then write

$$\underline{r}_n = \hat{\underline{g}}_{n+1} - \underline{g}_{n+1}$$

$$= \sum_{m=0}^{n} D_{nm} \, \underline{\varepsilon}_m \; , \qquad (B1)$$

where D_{nm} is a $(K+1) \times (K+1)$ matrix that determines the contribution of the local error $\underline{\varepsilon}_m$ to $\underline{r}_n$. The accumulated error $\underline{r}_n$ may also be written as

$$\underline{r}_n = J_n \, \underline{r}_{n-1} + \underline{\varepsilon}_n$$

$$= J_n \sum_{m=0}^{n-1} D_{n-1\,m} \, \underline{\varepsilon}_m + \underline{\varepsilon}_n \; . \qquad (B2)$$

Comparing the coefficients of $\{\underline{\varepsilon}_m\}$ in (B1) and (B2) establishes that

$$D_{nn} = I$$

and

$$D_{nj} = J_n \, D_{n-1\,j} = \prod_{i=j+1}^{n} J_i \; , \quad \begin{array}{l} j = 0,1,2,\ldots,n-1 \; , \\ n = 1,2,3,\ldots \quad . \end{array}$$

This leads to

$$\underline{r}_n = \prod_{i=1}^{n} J_i \, \underline{\varepsilon}_0 + \prod_{i=2}^{n} J_i \, \underline{\varepsilon}_1 + \ldots + J_n \, \underline{\varepsilon}_{n-1} + \underline{\varepsilon}_n \; . \qquad (B3)$$

A similar expression for recursive relations involving a single variable is given by Henrici [1962]. The application of (B3) will now be illustrated by considering a specific example.

Consider algorithm (3.4). Accordingly we have

$$
\left.
\begin{aligned}
g_{n+1,K} &= f_{n+1,0}\left(g_{n,K} + g_{n,K-1}\right) \\[2mm]
g_{n+1,j} &= f_{n+1,K-j}\left(g_{n,K} + g_{n,K-1}\right) + \sum_{i=0}^{K-j-1} f_{n+1,i}\, g_{n,i+j-1}\,, \\
& \qquad\qquad\qquad\qquad j = 1,2,3,\ldots,K-1 \\[2mm]
g_{n+1,0} &= 1 - \sum_{j=1}^{K} g_{n+1,j}
\end{aligned}
\right\} . \quad (B4)
$$

This leads to

$$
J_n =
\begin{bmatrix}
-f_{n+1,0} & -f_{n+1,0}-f_{n+1,1} & -f_{n+1,0}-f_{n+1,1}-f_{n+1,2} & \cdots & -\sum_{i=0}^{K-2} f_{n+1,i} & -\sum_{i=0}^{K-1} f_{n+1,i} & -\sum_{i=0}^{K-1} f_{n+1,i} \\
f_{n+1,0} & f_{n+1,1} & f_{n+1,2} & \cdots & f_{n+1,K-2} & f_{n+1,K-1} & f_{n+1,K-1} \\
0 & f_{n+1,0} & f_{n+1,1} & \cdots & f_{n+1,K-3} & f_{n+1,K-2} & f_{n+1,K-2} \\
0 & 0 & f_{n+1,0} & \cdots & \cdot & \cdot & \cdot \\
\cdot & \cdot & 0 & \cdots & \cdot & \cdot & \cdot \\
\cdot & \cdot & \cdot & \cdots & \cdot & \cdot & \cdot \\
\cdot & \cdot & \cdot & \cdots & f_{n+1,1} & f_{n+1,2} & f_{n+1,2} \\
\cdot & \cdot & \cdot & \cdots & f_{n+1,0} & f_{n+1,1} & f_{n+1,1} \\
\cdot & \cdot & \cdot & \cdots & 0 & f_{n+1,0} & f_{n+1,0}
\end{bmatrix}
$$

Since all the computations are carried out in floating point mode, one may assume that a single operation may introduce a maximum error ε, with $|\varepsilon| \leq 1.0 \times 10^{-t}$, when t decimal digits are retained by the machine. Also, it is known that in floating point operations (Fox and Mayers, [1968]),

$$\text{float } (X + Y) = (X + Y)(1 + \varepsilon)$$

and

$$\text{float } (XY) = XY(1 + \varepsilon) \ .$$

Application of these to (B4) leads to

$$\text{float } \left(g_{n+1,K}\right)$$

$$= f_{n+1,0}\left(g_{n,K} + g_{n,K-1}\right)(1+\varepsilon)^2 \ ,$$

$$\text{float}\left(g_{n+1,j}\right)$$

$$= \left[\left[f_{n+1,K-j}\left(g_{n,K} + g_{n,K-1}\right)(1+\varepsilon)^2 + f_{n+1,0}\ g_{n,j-1}(1+\varepsilon)\right](1+\varepsilon)\right.$$

$$\left. + f_{n+1,1}\ g_{n,j}(1+\varepsilon)\right]\left(1 + E_2\right) + f_{n+1,2}\ g_{n,j+1}(1+\varepsilon)\left(1 + E_3\right)$$

$$+ \cdots + f_{n+1,K-j-1}\ g_{n,K-2}(1+\varepsilon)\left(1 + E_{K-j}\right) \ ,$$

with $1 + E_i = (1+\varepsilon)^{K-j-i+1}$ and $j = 1,2,3,\ldots,K-1$. It is reasonable to assume that $(1+\varepsilon)^i = 1 + i\varepsilon$, for in most cases ε is quite small. As a result, it may be shown that the local error $\underline{\varepsilon}_n$ in the nth execution of the recursion formula equals

$$\left(\varepsilon_{n,0}\ \varepsilon_{n,1}\ \varepsilon_{n,2}\ \cdots\ \varepsilon_{n,K}\right)' \ ,$$

ILLUSTRATION B1:   THE JACOBIAN

-0.33333	-0.555556	-0.703704	-0.802469	-0.868313	-0.912209	-0.941472	-0.969982	-0.973908	-0.982658	-0.982658
0.33333	0.222222	0.148148	0.098765	0.065844	0.043896	0.029264	0.019509	0.013006	0.008671	0.008671
0.0	0.33333	0.222222	0.148148	0.098765	0.065844	0.043896	0.029264	0.019509	0.013006	0.013006
0.0	0.0	0.33333	0.222222	0.148148	0.098765	0.065844	0.043896	0.029264	0.019509	0.019509
0.0	0.0	0.0	0.33333	0.222222	0.148148	0.098765	0.065844	0.043896	0.029264	0.029264
0.0	0.0	0.0	0.0	0.33333	0.222222	0.148148	0.098765	0.065844	0.043896	0.043876
0.0	0.0	0.0	0.0	0.0	0.33333	0.222222	0.148148	0.098765	0.065844	0.065844
0.0	0.0	0.0	0.0	0.0	0.0	0.33333	0.222222	0.148148	0.098765	0.098765
0.0	0.0	0.0	0.0	0.0	0.0	0.0	0.33333	0.222222	0.148148	0.148148
0.0	0.0	0.0	0.0	0.0	0.0	0.0	0.0	0.33333	0.222222	0.222222
0.0	0.0	0.0	0.0	0.0	0.0	0.0	0.0	0.0	0.33333	0.33333

ILLUSTRATION B2: THE LOCAL ERROR COEFFICIENTS

n	α_0	α_1	α_2	α_3	α_4	α_5	α_6	α_7	α_8	α_9	α_{10}
1	0.0	0.0	0.0	0.0	0.0	0.0	0.0	0.0	0.0	0.0	0.0
2	-1.000000	1.000000	0.0	0.0	0.0	0.0	0.0	0.0	0.0	0.0	0.0
3	-1.382716	1.086420	0.296296	0.0	0.0	0.0	0.0	0.0	0.0	0.0	0.0
4	-1.570645	1.097394	0.386831	0.086420	0.0	0.0	0.0	0.0	0.0	0.0	0.0
5	-1.675659	1.094955	0.424326	0.131687	0.024691	0.0	0.0	0.0	0.0	0.0	0.0
6	-1.739217	1.090078	0.442615	0.156683	0.042981	0.006859	0.0	0.0	0.0	0.0	0.0
7	-1.779833	1.085291	0.452506	0.171519	0.055276	0.013413	0.001829	0.0	0.0	0.0	0.0
8	-1.806834	1.081169	0.458246	0.180852	0.063586	0.018561	0.003963	0.000457	0.0	0.0	0.0
9	-1.825330	1.077768	0.461754	0.186997	0.069326	0.022430	0.005871	0.001084	0.000102	0.0	0.0
10	-1.838462	1.075006	0.463989	0.191199	0.073395	0.025330	0.007458	0.001729	0.000282	0.000040	0.000034
11	-1.848051	1.072776	0.465461	0.194161	0.076347	0.027521	0.008745	0.002325	0.000499	0.000113	0.000102
12	-1.855212	1.070975	0.466458	0.196302	0.078533	0.029195	0.009781	0.002850	0.000723	0.000205	0.000190
13	-1.860657	1.069520	0.467149	0.197883	0.080179	0.030488	0.010614	0.003300	0.000935	0.000303	0.000287
14	-1.864856	1.068340	0.467638	0.199071	0.081437	0.031497	0.011284	0.003681	0.001127	0.000398	0.000382
15	-1.868134	1.067383	0.467992	0.199978	0.082412	0.032291	0.011824	0.004000	0.001296	0.000486	0.000471
16	-1.870718	1.066605	0.468252	0.200679	0.083175	0.032921	0.012262	0.004265	0.001443	0.000564	0.000551
17	-1.872770	1.065972	0.468447	0.201227	0.083778	0.033425	0.012617	0.004485	0.001567	0.000632	0.000621
18	-1.874411	1.065456	0.468595	0.201660	0.084257	0.033829	0.012905	0.004667	0.001672	0.000691	0.000681
19	-1.875730	1.065034	0.468709	0.202004	0.084641	0.034154	0.013140	0.004816	0.001759	0.000740	0.000731
20	-1.876793	1.064690	0.468798	0.202279	0.084950	0.034418	0.013331	0.004939	0.001832	0.000781	0.000774

ILLUSTRATION B3: THE ACCUMULATED ROUND-OFF ERROR COEFFICIENTS

n	β_0	β_1	β_2	β_3	β_4	β_5	β_6	β_7	β_8	β_9	β_{10}
1	0.0	0.0	0.0	0.0	0.0	0.0	0.0	0.0	0.0	0.0	0.0
2	-1.000000	1.000000	0.0	0.0	0.0	0.0	0.0	0.0	0.0	0.0	0.0
3	-1.604938	0.975309	0.629630	0.0	0.0	0.0	0.0	0.0	0.0	0.0	0.0
4	-2.020576	0.872428	0.851852	0.296296	0.0	0.0	0.0	0.0	0.0	0.0	0.0
5	-2.324036	0.770767	0.948331	0.481481	0.123457	0.0	0.0	0.0	0.0	0.0	0.0
6	-2.553659	0.682857	0.993802	0.598080	0.230910	0.048011	0.0	0.0	0.0	0.0	0.0
7	-2.731558	0.609428	1.015541	0.674643	0.313062	0.101052	0.017833	0.0	0.0	0.0	0.0
8	-2.871710	0.548731	1.025368	0.726820	0.374768	0.148013	0.041610	0.006401	0.0	0.0	0.0
9	-2.983510	0.498678	1.028974	0.763462	0.421340	0.187041	0.065403	0.016376	0.002235	0.0	0.0
10	-3.073714	0.457380	1.029304	0.789833	0.456859	0.218795	0.086986	0.027500	0.006238	0.000785	0.000034
11	-3.147207	0.423198	1.027951	0.809207	0.484263	0.244478	0.105751	0.038436	0.011173	0.002374	0.000374
12	-3.207579	0.394752	1.025786	0.823696	0.505648	0.265264	0.121753	0.048569	0.016425	0.004540	0.001106
13	-3.257516	0.371090	1.023294	0.834704	0.522518	0.282148	0.135281	0.057669	0.021611	0.007033	0.002169
14	-3.299058	0.351239	1.020745	0.843186	0.535963	0.295933	0.146679	0.065700	0.026516	0.009647	0.003450
15	-3.333776	0.334562	1.018289	0.849804	0.546779	0.307248	0.156279	0.072715	0.031029	0.012235	0.004837
16	-3.362899	0.320515	1.016005	0.855027	0.555552	0.316585	0.164368	0.078800	0.035105	0.014701	0.006242
17	-3.387395	0.308659	1.013927	0.859191	0.562720	0.324325	0.171192	0.084055	0.038737	0.016988	0.007602
18	-3.408044	0.298639	1.012068	0.862540	0.568613	0.330768	0.176954	0.088577	0.041941	0.019067	0.008877
19	-3.425474	0.290162	1.010422	0.865255	0.573433	0.336148	0.181824	0.092458	0.044745	0.020930	0.010046
20	-3.440201	0.282987	1.008979	0.867470	0.577525	0.340653	0.185942	0.095782	0.047184	0.022580	0.011100

where

$$\varepsilon_{n,0} = - \sum_{j=1}^{K} \varepsilon_{n,j} \; ,$$

$$\varepsilon_{n,j} = \left(g_{n,K} + g_{n,K-1}\right) f_{n+1,K-j} \, (K-j-2)\,\varepsilon$$

$$+ \sum_{i=0}^{K-j-1} f_{n+1,i} \; g_{n,j-1+i} \, (K-j-i+1)\,\varepsilon \; ,$$

$$j = 1,2,\ldots,K-1 \; ,$$

$$\varepsilon_{n,K} = \left(g_{n,K} + g_{n,K-1}\right) f_{n+1,0} \cdot 2\varepsilon \; .$$

(B5)

Bounds on the accumulated round-off error for a given number of recursions may now be obtained from (B3) and (B5). In order to demonstrate this, we consider the numerical example of Section 3.3. The Jacobians for this example are time-homogeneous, each being identical to the matrix given in Illustration B1. Computations to evaluate the successive local and accumulated round-off errors for this example, using (B3) and (B5), were accomplished by a FORTRAN IV program, listed at the end of this appendix. Illustrations B2 and B3 list the coefficients of the local and accumulated round-off errors respectively, for twenty successive recursions. These coefficients are defined as $\{\alpha_i, \; i = 0,1,2,\ldots,K\}$ and $\{\beta_i, \; i = 0,1,2,\ldots,K\}$ respectively, with

$$\text{local error} \quad \underline{\varepsilon}_n = \left(\alpha_0\varepsilon, \alpha_1\varepsilon, \alpha_2\varepsilon \; \cdots \; \alpha_K\varepsilon\right)'$$

and

$$\text{accumulated error} \quad \underline{r}_n = \left(\beta_0\varepsilon, \beta_1\varepsilon, \beta_2\varepsilon \; \cdots \; \beta_K\varepsilon\right)' \; ,$$

ε being as defined earlier. Illustration B4 lists the bounds on the

ILLUSTRATION B4: THE BOUNDS ON COMPUTATIONAL ERRORS

State i	State probability $g_{20,i}$	Maximum error in $g_{20,i}$	
		single precision	double precision
0	0.5014325	-3.440201×10^{-7}	$-3.440201 \times 10^{-16}$
1	0.2507162	0.282987×10^{-7}	0.282987×10^{-16}
2	0.1252236	1.008979×10^{-7}	1.008979×10^{-16}
3	0.0624100	0.865255×10^{-7}	0.865255×10^{-16}
4	0.0309853	0.577525×10^{-7}	0.577525×10^{-16}
5	0.0152876	0.340653×10^{-7}	0.340653×10^{-16}
6	0.0074715	0.185942×10^{-7}	0.185942×10^{-16}
7	0.0036034	0.095782×10^{-7}	0.095782×10^{-16}
8	0.0017087	0.047184×10^{-7}	0.047184×10^{-16}
9	0.0007955	0.022580×10^{-7}	0.022580×10^{-16}
10	0.0003657	0.011100×10^{-7}	0.011100×10^{-16}

accumulated errors existing at the end of twenty recursions, for the specific case where time-dependent state probabilities $\underline{g}_{20}$ $\left(= (g_{20,0} \ g_{20,1} \ \cdots \ g_{20,K})' \right)$ are computed on an IBM 370/165. The bounds specify the respective maximum possible errors in the computation of the state probabilities when single and double precision modes are employed. It is apparent that at least for the particular case considered here, algorithm (3.4) is numerically stable.

The statistical approach for analysing the errors suggested by Henrici regards the local errors as random variables. Since the local errors are independent of each other, it is possible to obtain an expression for the variance of the accumulated round-off error, making appropriate assumptions for the variance of the local errors. Such expressions would be particularly useful in making a probabilistic, rather than a deterministic "upper bound" type statement, about the computed results, when the upper bounds turn out to be uncomfortably large.

```
C
C     ERROR ANALYSER FOR ALGORITHM 3.4
C
      DIMENSION P(15,15),Q(15,15),F(11),A(11,11),
     1AJ(25,11,11),Y(25,11),X(11),G(25,11),E(11)
      DOUBLE PRECISION P,Q,R,AL,AMU,AMLM,ALLM,
     1F,X,Y,E,G,A,AJ,S,SUM,AKJ1,AKJ11
C
C     SET FORTH THE JACOBIAN AND OBTAIN ITS POWERINGS
C
      AL=1.0
      AMU=2.0
      KK=11
      AMLM=AMU/(AL+AMU)
      ALLM=AL/(AL+AMU)
      DO 99 I=1,KK
      J=I-1
      F(I)=ALLM*AMLM**J
99    CONTINUE
      WRITE(6,150)
      DO 35 I=1,KK
      DO 35 J=1,KK
      P(I,J)=0.0
35    CONTINUE
      K1=KK-1
      DO 31 I=2,KK
      I1=I-1
      L=1
      DO 30 J=I1,K1
      P(I,J)=F(L)
      L=L+1
30    CONTINUE
      P(I,KK)=P(I,KK-1)
31    CONTINUE
      P(1,1)=-F(1)
      KKK=KK-1
      DO 32 I=2,KKK
      P(1,I)=P(1,I-1)-F(I)
32    CONTINUE
      P(1,KK)=P(1,KK-1)
102   FORMAT(50X,7HPOWER =,I6)
      DO 4 I=1,KK
101   FORMAT(11F11.6,F12.6)
4     CONTINUE
      N=1
      DO 9 I=1,KK
      DO 9 J=1,KK
      AJ(1,I,J)=P(I,J)
      A(I,J)=P(I,J)
9     CONTINUE
5     CONTINUE
      N=N+1
100   FORMAT(////////)
      DO 6 I=1,KK
      DO 7 J=1,KK
      Q(I,J)=0.0
      DO 3 L=1,KK
      Q(I,J)=Q(I,J)+A(I,L)*P(L,J)
```

```
0051        8      CONTINUE
0052        7      CONTINUE
0053        6      CONTINUE
0054               DO 21 I=1,KK
0055               DO 22 J=1,KK
0056               A(I,J)=Q(I,J)
0057               AJ(N,I,J)=Q(I,J)
0058       22      CONTINUE
0059               R=0.0
0060               DO 25 J=1,KK
0061               R=R+A(I,J)
0062       25      CONTINUE
0063       21      CONTINUE
0064               IF(N.GT.20) GO TO 23
0065               GO TO 5
0066       23      CONTINUE
           C
0067               K=11
0068               DO 10 I=1,K
0069               G(1,I)=0.0
0070       10      CONTINUE
           C
           C       INITIALIZE THE QUEUING SYSTEM AND GENERATE
           C       THE TIME-DEPENDENT STATE PROBABILITIES
           C
0071               G(1,1)=1.0
0072               DO 11 I=1,K
0073               J=I-1
0074               F(I)=ALLM*AMLM**J
0075       11      CONTINUE
0076               WRITE(6,100)
0077               DO 17 M=1,20
0078               DO 307 N=1,20
0079               DO 307 I=1,11
0080               X(I)=0.0
0081               Y(N,I)=0.0
0082      307      CONTINUE
0083               DO 15 N=1,M
0084               G(N+1,K)=F(1)*(G(N,K)+G(N,K-1))
0085               K2=K-2
0086               DO 14 J=1,K2
0087               SUM=0.0
0088               DO 12 I=1,J
0089               SUM=SUM+F(I)*G(N,K-J+I-2)
0090       12      CONTINUE
0091               G(N+1,K-J)=F(J+1)*(G(N,K)+G(N,K-1))+SUM
0092       14      CONTINUE
0093               S=0.0
0094               DO 16 J=2,K
0095               S=S+G(N+1,J)
0096       16      CONTINUE
0097               G(N+1,1)=1.0-S
           C
           C       COMPUTE THE LOCAL ERROR COEFFICIENTS
           C
0098               KK=K
0099               E(KK)=(G(N,KK-1))*F(1)*2.0
0100               KK1=KK-1
```

```
        DO 1 J=2,KK1
        AKJ1=KK-J+1
        E(J)=F(KK-J+1)*(G(N,KK)+G(N,KK-1))*AKJ1
        KJM1=KK-J
        DO 1 I=1,KJM1
        AKJ11=KK-J+1-I
        E(J)=E(J)+F(I)*G(N,J-1+I)*AKJ11
1       CONTINUE
        E(1)=-E(2)-E(3)-E(4)-E(5)-E(6)-E(7)-E(8)-E(9)-E(10)-E(11)
C
C       COMPUTE  THE  ACCUMULATED  ERROR  CCEFFICIENTS
C
        IF(N.EQ.M) GC TO 305
        DO 300  I=1,KK
        DO 300  J=1,KK
        Y(N,I)=Y(N,I)+AJ(M-N,I,J)*E(J)
300     CONTINUE
        DO 301 I=1,KK
        X(I)=X(I)+Y(N,I)
301     CONTINUE
103     FORMAT(/I10,F12.6,10F9.6)
        GO TO 15
305     CONTINUE
        DO 302 I=1,KK
        Y(N,I)=Y(N,I)+E(I)
302     CONTINUE
        DO 303 I=1,KK
        X(I)=X(I)+Y(N,I)
303     CONTINUE
15      CONTINUE
        WRITE(6,103) N,(X(I),I=1,11)
17      CONTINUE
        WRITE(6,150)
150     FORMAT(1H1)
        STOP
        END
```

APPENDIX C

ON ANALYTIC APPROXIMATIONS

SOME RESULTS IN APPROXIMATION.   We saw in Chapter III that accurate
numerical results may be obtained for bulk queuing problems by employing the concept of sweeping of probabilities. However, some inequalities for the expected queue size in a stationary bulk queue are easy
to evaluate, and the accuracy of bounds on the expected queue size
established by utilizing these inequalities may sometimes be sufficient
for the purpose of applications to practical problems. Some such
methods for an infinite waiting space bulk queue are considered in
this appendix.

Let $\{w_n\}$ be a string of random variables, satisfying the Lindley
equation

$$w_{n+1} = \left(w_n + u_n\right)^+ ,$$

where for any real number x , $x^+ = \max(x, 0)$ and $\{u_n\}$ is a string
of independent and identically distributed (i.i.d.) random variables.
It can be shown that $w_n \to w$ as $n \to \infty$ when $E\left(u_n\right) < 0$ (see Lindley
[1952]). Kingman [1962a] has given some analytic inequalities, stated
below, which pertain to the stationary expected value of w_n , $E(w)$.
These inequalities may also be applied to examine the queue length
process of the bulk queuing problem.

Consider a simple queuing system in which customers C_0 , C_1 ,
C_2 , $\dots$ arrive and are attended to by a single server in the order
of their arrival. In Kingman's notation, the customer C_n waits in

the queue for a time w_n , and is served in a further time s_n . The time elapsing between the arrivals of successive customers C_n , C_{n+1} is denoted by t_n . Also,

$$u_n = s_n - t_n , \qquad U_n = \sum_{r=0}^{n-1} u_r .$$

It is supposed that C_0 finds the queue empty and the server free, implying thereby that $w_0 = 0$. The waiting times of the successive customers are related by the Lindley equation. The random variables $\{u_n\}$ are i.i.d. by assumption. If these variables are distributed as u , and $\{s_n\}$ and $\{t_n\}$ are distributed as s and t respectively, then it may be shown that

1) If $E(s) < E(t) < \infty$, and if, for some $\varepsilon > 0$

$$E\left(e^{\varepsilon s}\right) < \infty ,$$

then for all $\theta > 0$ such that $E\left(e^{\theta u}\right) < 1$,

$$E(w) \leq - (e\theta)^{-1} \ell n\left[1 - E\left(e^{\theta u}\right)\right] . \tag{C1}$$

There is a value of θ for which the right hand side of (C1) attains its least value (Theorem 3, page 319, Kingman [1962a]).

2) If $E\left(s^2\right)$, $E\left(t^2\right) < \infty$, and if $E(u) < 0$, then

$$E(w) \leq \tfrac{1}{2} \operatorname{Var}(u)/E(-u) . \tag{C2}$$

The inequality approaches equality asymptotically in conditions of heavy traffic (Theorem 2, page 316, Kingman [1962a]).

The above results apply to the stationary waiting time in the GI/G/1 queue. These results cannot be applied to the general bulk queue $GI^X/G^Y/1$, since the waiting time process is greatly compli- cated by bulk arrival and service. It is possible, however, to apply these in a simple way to study the queue length process in the $GI^X/M^Y/1$ or the $M^X/G^Y/1$ queue. This is possible because the random variables $\{X_n\}$ and $\{Y_n\}$, defined while constructing the imbedded Markov chains for the last two queues (Processes A and B, Chapter II), are (respectively) independent and identically distributed random variables. We consider below the $GI^X/M^Y/1$ queuing system.

UPPER AND LOWER BOUNDS FOR EXPECTED QUEUE SIZE. By defining the sets of random variables $\{X_n\}$ and $\{Y_{n+1}\}$ appropriately, the $GI^X/M^Y/1$ queue may be modeled as a Markov chain (see Process B, Section 2.1). Thus one obtains

$$Q_{n+1} = (Q_n + X_n - Y_{n+1})^+ , \quad n = 0,1,2,\ldots . \tag{2.2}$$

Since the sets $\{X_n\}$ and $\{Y_n\}$ are independent of each other, a new sequence of independent identically distributed random variables may be constructed as $(X_n - Y_{n+1})$, $n = 0,1,2,\ldots$. A simple trans- formation of Kingman's results is obtained, if the random variable $(X_n - Y_{n+1})$ is identified with u_n (or u). This enables the deriva- tion of upper bounds for the expected queue size $E(Q)$ for a stationary $GI^X/M^Y/1$ queuing system to be made.

Let $s = e^\theta$, when θ is a positive number. If in addition $(X_n - Y_{n+1})$ is identified with u , then it may be shown that

$$E(Q) \leq \min_{s(s>1)} \frac{1}{e \ln s} \ln \left(\frac{1}{1 - E\left(s^{X_n - Y_{n+1}}\right)} \right) \tag{C3}$$

where it is assumed that $Q_n \to Q$ as $n \to \infty$. The right hand side of

(C3) may be evaluated either analytically or numerically. We shall let

E_1 denote the upper bound for $E(Q)$ that is evaluated on the basis of

(C3). Inequality (C2) above may be utilized to derive a different

upper bound for $E(Q)$ that becomes tight in the heavy traffic situa-

tion. Accordingly, we have

$$E(Q) \leq \tfrac{1}{2} \frac{\mathrm{Var}\left(X_n - Y_{n+1}\right)}{E\left(Y_{n+1} - X_n\right)} . \qquad (C4)$$

We shall let E_2 denote the upper bound for $E(Q)$, based on (C4).

In practice, it would be of interest to evaluate both E_1 and E_2 :

The smaller of the two bounds being the one to be accepted as an

"upper bound" for $E(Q)$. This procedure is illustrated in the follow-

ing example.

The Markovian single server bulk queue $M^X/M^Y/1$ may be readily

specialized from the results for the $GI^X/M^Y/1$, where the exponen-

tial distribution replaces the general distribution of interarrival

times. Assume that the sizes of the arriving batches are distributed

geometrically, with $\mathrm{Prob}[\text{batch size} = v] = (1-p)p^{v-1}$, $p < 1$ and

$v = 1,2,3,\ldots$. The capacity of the server to accept a batch during

a given service period is a random variable, also distributed geo-

metrically, with $\mathrm{Prob}[\text{capacity} = v] = (1-q)q^{v-1}$, $q < 1$ and

$v = 1,2,3,\ldots$. To ensure stationarity, also assume that

$\rho = \lambda(1-q)/\left(\mu(1-p)\right) < 1$ (Theorem 8, Bhat [1964]).

It may be shown that

$$E\left(s^{X_n}\right) = (1-p)s/(1-ps) .$$

Also, $\text{Prob}[Y_{n+1}=j] = \lambda(a-q)a^{j-1}\big/(\lambda+\mu)$, $j = 1,2,3,\ldots$, with $a = (\mu+\lambda q)/(\lambda+\mu)$ and $\text{Prob}[Y_{n+1}=0] = \lambda/(\lambda+\mu)$. Hence

$$E\left(s^{Y_{n+1}}\right) = \lambda(1-qs)/\big((\lambda+\mu)(1-as)\big)$$

and

$$E\left(s^{(X_n-Y_{n+1})}\right) = \frac{(1-p)\,s\lambda\,(s-q)}{(1-ps)\,(s\lambda+s\mu-\mu-\lambda q)} \quad .$$

$E(X_n)$, $E(Y_{n+1})$, $\text{Var}(X_n)$ and $\text{Var}(Y_{n+1})$ may be obtained from the above, and hence both E_1 and E_2 can be found. A numerical problem is considered in the sequel.

A lower bound for $E(Q)$ might also be given. From (2.2) it follows that Q_n has the same distribution as $\tilde{Q}_n$ with

$$\tilde{Q}_n = \max_{0\le m\le n} \sum_{r=1}^{m} (X_r - Y_{r+1})$$

since $\{X_r - Y_{r+1}\}$ is a sequence of independent identically distributed random variables. When the bulk queue is stable, the distribution of $\tilde{Q}_n$ approaches the stationary queue size distribution. Also, it can be shown that

$$E(Q_n) = \sum_{r=1}^{n} \frac{1}{r} E(S_r^+)$$

where $S_r = \sum_{i=0}^{r-1} (X_i - Y_{i+1})$ and $X^+ = \max(X, 0)$ for any real number X . The proofs here are implicit in the work of Kingman [1962a] and follow immediately from appropriate identification of the random variables. Assume that $Q_0 = 0$. From the monotonicity properties of the distribution of Q_n it would follow that

$$E(Q) \geq E(Q_n)$$

(cf. Kingman [1962a], Theorem 4), and as a consequence $E(Q_n)$ (for some value of n) might be treated as a lower bound for $E(Q)$. This bound would be asymptotically sharp from below, with increase in the value of n.

A numerical evaluation of $E(s_r^+)$, for arbitrary value of r, however, presents considerable difficulty. A more recent result of Kingman may be utilized to evaluate a somewhat less sharp lower bound for $E(Q)$. Kingman [1970] shows that

$$E(w) \geq \frac{E\{(u^+)^2\}}{2E(-u)} .$$

As before, if $(X_n - Y_{n+1})$ in (2.2) is identified with u, then this result may be used to obtain a lower bound for $E(Q)$, when Q is identified with w. Accordingly,

$$E(Q) \geq \frac{E\left\{[(X_n - Y_{n+1})^+]^2\right\}}{2E(Y_{n+1} - X_n)} . \tag{C5}$$

Consider, for example, the Markovian bulk queue described earlier. It can be shown that

$$E\left(s^{(X_n - Y_{n+1})^+}\right) = \frac{1-p}{1 - pa}\left[\frac{\mu}{\mu+\lambda} + \frac{\lambda}{\mu+\lambda}\frac{s(1 - pq)}{1 - ps}\right] ,$$

where $a = (\mu + \lambda q)/(\mu+\lambda)$. From this one obtains

$$E\left\{\left[\left(X_n - Y_{n+1}\right)^+\right]^2\right\} = \frac{\lambda}{\mu+\lambda} \frac{(1+p)(1-pq)}{(1-pa)(1-p)^2} \quad .$$

The lower bound for $E(Q)$ may therefore be found from (C5).

Table C1 displays the upper and lower bounds for $E(Q)$ in a numerical problem, where $\mu = 1.0$, $p = 0.5$, $q = 0.5$ and $\lambda = \rho$.

TABLE C1:   UPPER AND LOWER BOUNDS FOR EXPECTED QUEUE SIZE

Traffic Intensity ρ	E_1	E_2	Upper Bound	Lower Bound
0.01	0.055	102.5	0.055	0.000224
0.05	0.289	22.7	0.289	0.00551
0.1	0.612	12.8	0.612	0.0217
0.2	1.41	8.25	1.41	0.0865
0.3	2.45	7.12	2.45	0.200
0.4	4.20	7.00	4.20	0.375
0.5	6.23	7.50	6.23	0.643
0.6	9.94	8.67	8.67	1.07
0.7	16.8	10.93	10.93	1.79
0.8	...	15.75	15.75	3.27
0.9	...	30.61	30.61	7.75
0.95	...	60.53	60.53	16.75
0.99	...	300.5	300.5	88.74

The computational times required to evaluate both the upper and the lower bounds, tabulated in Table C1, were of the order of a few seconds, when an IBM 370/165 system was employed. Nevertheless, as a comparison of the values of upper and lower bounds thus obtained would suggest, a set of tighter pair of bounds would perhaps be desirable.

The methods employed to obtain the upper bound yield comparatively sharper results when compared with the corresponding formula (C5), which yields a lower bound. In order to obtain a sharper set of lower bounds, two avenues would seem worthwhile to pursue. First, attempts should be directed toward developing a simple method for numerically evaluating $E\left(S_r^+\right)$. Alternatively a numerical solution to a discrete version of Marshall's [1968] equation (14), which would involve the distribution of $\left(X_n - Y_{n+1}\right)$, could be attempted.

Lecture Notes in Economics and Mathematical Systems

(Vol. 1–15: Lecture Notes in Operations Research and Mathematical Economics, Vol. 16–59: Lecture Notes in Operations Research and Mathematical Systems)

Vol. 1: H. Bühlmann, H. Loeffel, E. Nievergelt, Einführung in die Theorie und Praxis der Entscheidung bei Unsicherheit. 2. Auflage, IV, 125 Seiten 4°. 1969. DM 16,–

Vol. 2: U. N. Bhat, A Study of the Queueing Systems M/G/1 and GI/M/1. VIII, 78 pages. 4°. 1968. DM 16,–

Vol. 3: A. Strauss, An Introduction to Optimal Control Theory. VI, 153 pages. 4°. 1968. DM 16,–

Vol. 4: Einführung in die Methode Branch and Bound. Herausgegeben von F. Weinberg. VIII, 159 Seiten. 4°. 1968. DM 16,–

Vol. 5: Hyvärinen, Information Theory for Systems Engineers. VIII, 205 pages. 4°. 1968. DM 16,–

Vol. 6: H. P. Künzi, O. Müller, E. Nievergelt, Einführungskursus in die dynamische Programmierung. IV, 103 Seiten. 4°. 1968. DM 16,–

Vol. 7: W. Popp, Einführung in die Theorie der Lagerhaltung. VI, 173 Seiten. 4°. 1968. DM 16,–

Vol. 8: J. Teghem, J. Loris-Teghem, J. P. Lambotte, Modèles d'Attente M/G/1 et GI/M/1 à Arrivées et Services en Groupes. IV, 53 pages. 4°. 1969. DM 16,–

Vol. 9: E. Schultze, Einführung in die mathematischen Grundlagen der Informationstheorie. VI, 116 Seiten. 4°. 1969. DM 16,–

Vol. 10: D. Hochstädter, Stochastische Lagerhaltungsmodelle. VI, 269 Seiten. 4°. 1969. DM 18,–

Vol. 11/12: Mathematical Systems Theory and Economics. Edited by H. W. Kuhn and G. P. Szegö. VIII, IV, 486 pages. 4°. 1969. DM 34,–

Vol. 13: Heuristische Planungsmethoden. Herausgegeben von F. Weinberg und C. A. Zehnder. II, 93 Seiten. 4°. 1969. DM 16,–

Vol. 14: Computing Methods in Optimization Problems. Edited by A. V. Balakrishnan. V, 191 pages. 4°. 1969. DM 16,–

Vol. 15: Economic Models, Estimation and Risk Programming: Essays in Honor of Gerhard Tintner. Edited by K. A. Fox, G. V. L. Narasimham and J. K. Sengupta. VIII, 461 pages. 4°. 1969. DM 24,–

Vol. 16: H. P. Künzi und W. Oettli, Nichtlineare Optimierung: Neuere Verfahren, Bibliographie. IV, 180 Seiten. 4°. 1969. DM 16,–

Vol. 17: H. Bauer und K. Neumann, Berechnung optimaler Steuerungen, Maximumprinzip und dynamische Optimierung. VIII, 188 Seiten. 4°. 1969. DM 16,–

Vol. 18: M. Wolff, Optimale Instandhaltungspolitiken in einfachen Systemen. V, 143 Seiten. 4°. 1970. DM 16,–

Vol. 19: L. Hyvärinen, Mathematical Modeling for Industrial Processes. VI, 122 pages. 4°. 1970. DM 16,–

Vol. 20: G. Uebe, Optimale Fahrpläne. IX, 161 Seiten. 4°. 1970. DM 16,–

Vol. 21: Th. Liebling, Graphentheorie in Planungs- und Tourenproblemen am Beispiel des städtischen Straßendienstes. IX, 118 Seiten. 4°. 1970. DM 16,–

Vol. 22: W. Eichhorn, Theorie der homogenen Produktionsfunktion. VIII, 119 Seiten. 4°. 1970. DM 16,–

Vol. 23: A. Ghosal, Some Aspects of Queueing and Storage Systems. IV, 93 pages. 4°. 1970. DM 16,–

Vol. 24: Feichtinger, Lernprozesse in stochastischen Automaten. V, 66 Seiten. 4°. 1970. DM 16,–

Vol. 25: R. Henn und O. Opitz, Konsum- und Produktionstheorie. I. II, 124 Seiten. 4°. 1970. DM 16,–

Vol. 26: D. Hochstädter und G. Uebe, Ökonometrische Methoden. XII, 250 Seiten. 4°. 1970. DM 18,–

Vol. 27: I. H. Mufti, Computational Methods in Optimal Control Problems. IV, 45 pages. 4°. 1970. DM 16,–

Vol. 28: Theoretical Approaches to Non-Numerical Problem Solving. Edited by R. B. Banerji and M. D. Mesarovic. VI, 466 pages. 4°. 1970. DM 24,–

Vol. 29: S. E. Elmaghraby, Some Network Models in Management Science. III, 177 pages. 4°. 1970. DM 16,–

Vol. 30: H. Noltemeier, Sensitivitätsanalyse bei diskreten linearen Optimierungsproblemen. VI, 102 Seiten. 4°. 1970. DM 16,–

Vol. 31: M. Kühlmeyer, Die nichtzentrale t-Verteilung. II, 106 Seiten. 4°. 1970. DM 16,–

Vol. 32: F. Bartholomes und G. Hotz, Homomorphismen und Reduktionen linearer Sprachen. XII, 143 Seiten. 4°. 1970. DM 16,–

Vol. 33: K. Hinderer, Foundations of Non-stationary Dynamic Programming with Discrete Time Parameter. VI, 160 pages. 4°. 1970. DM 16,–

Vol. 34: H. Störmer, Semi-Markoff-Prozesse mit endlich vielen Zuständen. Theorie und Anwendungen. VII, 128 Seiten. 4°. 1970. DM 16,–

Vol. 35: F. Ferschl, Markovketten. VI, 168 Seiten. 4°. 1970. DM 16,–

Vol. 36: M. P. J. Magill, On a General Economic Theory of Motion. VI, 95 pages. 4°. 1970. DM 16,–

Vol. 37: H. Müller-Merbach, On Round-Off Errors in Linear Programming. VI, 48 pages. 4°. 1970. DM 16,–

Vol. 38: Statistische Methoden I, herausgegeben von E. Walter. VIII, 338 Seiten. 4°. 1970. DM 22,–

Vol. 39: Statistische Methoden II, herausgegeben von E. Walter. IV, 155 Seiten. 4°. 1970. DM 16,–

Vol. 40: H. Drygas, The Coordinate-Free Approach to Gauss-Markov Estimation. VIII, 113 pages. 4°. 1970. DM 16,–

Vol. 41: U. Ueing, Zwei Lösungsmethoden für nichtkonvexe Programmierungsprobleme. VI, 92 Seiten. 4°. 1971. DM 16,–

Vol. 42: A. V. Balakrishnan, Introduction to Optimization Theory in a Hilbert Space. IV, 153 pages. 4°. 1971. DM 16,–

Vol. 43: J. A. Morales, Bayesian Full Information Structural Analysis. VI, 154 pages. 4°. 1971. DM 16,–

Vol. 44: G. Feichtinger, Stochastische Modelle demographischer Prozesse. XIII, 404 Seiten. 4°. 1971. DM 28,–

Vol. 45: K. Wendler, Hauptaustauschschritte (Principal Pivoting). II, 64 Seiten. 4°. 1971. DM 16,–

Vol. 46: C. Boucher, Leçons sur la théorie des automates mathématiques. VIII, 193 pages. 4°. 1971. DM 18,–

Vol. 47: H. A. Nour Eldin, Optimierung linearer Regelsysteme mit quadratischer Zielfunktion. VIII, 163 Seiten. 4°. 1971. DM 16,–

Vol. 48: M. Constam, Fortran für Anfänger. VI, 143 Seiten. 4°. 1971. DM 16,–

Vol. 49: Ch. Schneeweiß, Regelungstechnische stochastische Optimierungsverfahren. XI, 254 Seiten. 4°. 1971. DM 22,–

Vol. 50: Unternehmensforschung Heute – Übersichtsvorträge der Züricher Tagung von SVOR und DGU, September 1970. Herausgegeben von M. Beckmann. VI, 133 Seiten. 4°. 1971. DM 16,–

Vol. 51: Digitale Simulation. Herausgegeben von K. Bauknecht und W. Nef. IV, 207 Seiten. 4°. 1971. DM 18,–

Vol. 52: Invariant Imbedding. Proceedings of the Summer Workshop on Invariant Imbedding Held at the University of Southern California, June – August 1970. Edited by R. E. Bellman and E. D. Denman. IV, 148 pages. 4°. 1971. DM 16,–

Vol. 53: J. Rosenmüller, Kooperative Spiele und Märkte. IV, 152 Seiten. 4°. 1971. DM 16,–

Vol. 54: C. C. von Weizsäcker, Steady State Capital Theory. III, 102 pages. 4°. 1971. DM 16,–

Vol. 55: P. A. V. B. Swamy, Statistical Inference in Random Coefficient Regression Models. VIII, 209 pages. 4°. 1971. DM 20,–

Vol. 56: Mohamed A. El-Hodiri, Constrained Extrema. Introduction to the Differentiable Case with Economic Applications. III, 130 pages. 4°. 1971. DM 16,–

Vol. 57: E. Freund, Zeitvariable Mehrgrößensysteme. VII, 160 Seiten. 4°. 1971. DM 18,–

Vol. 58: P. B. Hagelschuer, Theorie der linearen Dekomposition. VII, 191 Seiten. 4°. 1971. DM 18,–

Vol. 59: J. A. Hanson, Growth in Open Economics. IV, 127 pages. 4°. 1971. DM 16,–

Vol. 60: H. Hauptmann, Schätz- und Kontrolltheorie in stetigen dynamischen Wirtschaftsmodellen. V, 104 Seiten. 4°. 1971. DM 16,–

Vol. 61: K. H. F. Meyer, Wartesysteme mit variabler Bearbeitungsrate. VII, 314 Seiten. 4°. 1971. DM 24,–

Vol. 62: W. Krelle u. G. Gabisch unter Mitarbeit von J. Burgermeister, Wachstumstheorie. VII, 223 Seiten. 4°. 1972. DM 20,–

Vol. 63: J. Kohlas, Monte Carlo Simulation im Operations Research. VI, 162 Seiten. 4°. 1972. DM 16,–

Vol. 64: P. Gessner u. K. Spremann, Optimierung in Funktionenräumen. IV, 120 Seiten. 4°. 1972. DM 16,–

Vol. 65: W. Everling, Exercises in Computer Systems Analysis. VIII, 184 pages. 4°. 1972. DM 18,–

Vol. 66: F. Bauer, P. Garabedian and D. Korn, Supercritical Wing Sections. V, 211 pages. 4°. 1972. DM 20,–

Vol. 67: I. V. Girsanov, Lectures on Mathematical Theory of Extremum Problems. V, 136 pages. 4°. 1972. DM 16,–

Vol. 68: J. Loeckx, Computability and Decidability. An Introduction for Students of Computer Science. VI, 76 pages. 4°. 1972. DM 16,–